SpringerBriefs in Computer Science

T0172043

SpringerBriefs present concise summaries of cutting-edge research and practical applications across a wide spectrum of fields. Featuring compact volumes of 50 to 125 pages, the series covers a range of content from professional to academic. Typical topics might include:

- A timely report of state-of-the art analytical techniques
- A bridge between new research results, as published in journal articles, and a contextual literature review
- A snapshot of a hot or emerging topic
- An in-depth case study or clinical example
- A presentation of core concepts that students must understand in order to make independent contributions

Briefs allow authors to present their ideas and readers to absorb them with minimal time investment. Briefs will be published as part of Springer's eBook collection, with millions of users worldwide. In addition, Briefs will be available for individual print and electronic purchase. Briefs are characterized by fast, global electronic dissemination, standard publishing contracts, easy-to-use manuscript preparation and formatting guidelines, and expedited production schedules. We aim for publication 8–12 weeks after acceptance. Both solicited and unsolicited manuscripts are considered for publication in this series.

More information about this series at http://www.springer.com/series/10028

Franklin de Lima Marquezino • Renato Portugal
Carlile Lavor

A Primer on Quantum Computing

 Springer

Franklin de Lima Marquezino
Federal University of Rio de Janeiro
Rio de Janeiro
Rio de Janeiro, Brazil

Renato Portugal
National Lab of Scientific Computing
Petrópolis
Rio de Janeiro, Brazil

Carlile Lavor
Department of Applied Mathematics
(IMECC-UNICAMP)
University of Campinas
Campinas
São Paulo, Brazil

ISSN 2191-5768 ISSN 2191-5776 (electronic)
SpringerBriefs in Computer Science
ISBN 978-3-030-19065-1 ISBN 978-3-030-19066-8 (eBook)
https://doi.org/10.1007/978-3-030-19066-8

This Springer imprint is published by the registered company Springer Nature Switzerland AG.
The registered company address is: Gewerbestrasse 11, 6330 Cham, Switzerland

Preface

Quantum computation is an exciting field which is attracting increasing attention unremittingly since the early 1980s. This field rests on the shoulder of a giant because classical computation achieved amazing developments in the twentieth century. It is exciting because quantum computation is a paradigm change. The basic information unit—the qubit—behaves quite differently from the bit, which is either zero or one exclusively. The qubit admits the coexistence of zero and one. It is not possible to visualize this behavior directly; in fact, mathematics is the key that has unlocked the quantum behavior and allowed physicists to develop quantum mechanics, which predicts experimental results that have been successfully tested in labs. A paradigm change means that a new building must be made with different foundations.

In the last years, we have learned that companies such as IBM, Google, and others are building quantum computers with more and more qubits. IBM has made available 5- and 14-qubit quantum computers for any user with a composer that is very easy to use—a mouse click holds a quantum gate, then you drag and drop. After playing with the mouse, you hit the run button, and, *voilà*, the magic happens. The first part of the magic comes from classical computation and its graphical interface; the last part comes from the *superconducting transmon qubits* working under temperature as low as one-third of deep space, a pinnacle of success combining cutting-edge technology and centuries of scientific development.

To date, the so-called quantum supremacy has not been achieved. Meanwhile, classical computers continue to speed up and, nowadays, are possible to simulate the behavior of 50 qubits approximately in supercomputers. Quantum computers with more than 50 qubits are candidates to achieve the supremacy. The problem is to maintain the coexistence of zeros and ones for long enough, let us say, seconds. IBM's devices are working up to 100 ms. It is a huge technological challenge to scale up the quantum behavior to macroscopic sizes that humans can see and time spans that humans can take a breath.

Of course, we can always say that *the time is now*. Yes, we confess we believe that the time is now. Feynman said in 1982 that quantum computing is an interesting research problem, and it still is. This motivates us to finish this project which we started over a decade and a half ago. We hope you enjoy it.

Petrópolis, Brazil   Franklin de Lima Marquezino
Petrópolis, Brazil   Renato Portugal
Campinas, Brazil   Carlile Lavor
March 2019

Acknowledgments

We are grateful to our colleagues and students from the Federal University of Rio de Janeiro (UFRJ, Brazil), the National Laboratory for Scientific Computing (LNCC, Brazil), and the University of Campinas (UNICAMP, Brazil) for several important discussions and interesting ideas.

We acknowledge CAPES, CNPq, FAPERJ, and FAPESP—Brazilian funding agencies—for the financial support to our research projects. We also thank the Brazilian Society of Computational and Applied Mathematics (SBMAC) for the opportunity to give a course on this subject that resulted in the first version of this monograph in Portuguese (http://www.sbmac.org.br/arquivos/notas/livro_08.pdf), which in turn evolved from our earliest tutorials (in arXiv quant-ph/0301079 and quant-ph/0303175).

Contents

Chapter 1
Introduction

"Where a calculator like ENIAC today is equipped with 18,000 vacuum tubes and weighs 30 tons, computers in the future may have only 1000 vacuum tubes and perhaps weigh only 1.5 tons."
(Popular Mechanics, March 1949)

In 1965, Gordon Moore observed that the number of basic components per integrated circuit roughly doubled every year, and predicted that this trend would remain constant for at least one decade [13]. Since then, this projection of exponential growth on the number of transistors in integrated circuits is known as the Moore's law. An important consequence of Moore's law is that in order to cram more transistors into the same area, their size should also decrease. In fact, technology advanced a lot since Moore's observations, to the point that nowadays it is possible to build a transistor with just a few nanometers, and pack hundreds of millions of them in each square millimeter of an integrated circuit.[1] This miniaturization process allows modern computers to perform more operations per unit of time while consuming less energy. However, there is a physical limit for this shrinkage. Moreover, if the basic components of integrated circuits become too small, their behavior must be described by the rules of quantum mechanics.

Almost 20 years after Gordon Moore's historical paper, another brilliant observation regarding the future of computers was presented. In 1982, Nobel Prize laureate Richard Feynman gave an inspired lecture entitled "Simulating Physics with Computers" [7], where he pointed that classical computers are inherently inefficient for simulating quantum systems in general, and proposed the revolutionary idea of quantum computing. Russian mathematician Yuri Manin independently came to similar conclusions in his book "Computable and Uncomputable" (in Russian), published in 1980 [12]. Roughly speaking, the reason for the aforementioned inefficiency is that the Hilbert space required to represent the quantum states grows exponentially with the number of particles being simulated, so the memory

---

[1]R. Courtland, "Intel Now Packs 100 Million Transistors in Each Square Millimeter", IEEE Spectrum, https://bit.ly/2yRyspF.

© The Author(s), under exclusive license to Springer Nature Switzerland AG 2019
F. de Lima Marquezino et al., *A Primer on Quantum Computing*, SpringerBriefs in
Computer Science, https://doi.org/10.1007/978-3-030-19066-8_1

necessary to represent only a few hundred of quantum particles is much beyond the capabilities of any classical computer. Feynman suggested that a computer made of quantum systems might be efficient for that task. Quantum mechanics simulation has many applications in fields such as chemistry, drug design, and nanotechnology, just to name a few.

Therefore, it is expected that a quantum computer should be more efficient than any classical computer in at least one task with several practical applications, namely the simulation of quantum systems. However, the idea that quantum computers can simulate quantum mechanics efficiently is not so surprising after all. On the other hand, being able to use quantum mechanics to solve different problems efficiently is indeed remarkable. The first hint that this would be possible came in 1985, with David Deutsch's paper about a quantum generalization of the universal Turing machine [4]. In 1992, some of those ideas were expanded and Deutsch-Jozsa algorithm was presented [5] as a quantum algorithm exponentially faster than any classical counterpart for a problem not directly related to quantum simulation—although rather artificial. In the 1990s, two quantum algorithms have been developed for very practical computational problems: Shor's algorithm for factoring large integers [15], and Grover's algorithm for searching on unstructured databases [8].

Shor's algorithm gives an exponential speedup compared to the best known classical algorithm, and has practical applications to code breaking. Grover's algorithm, on the other hand, has only quadratic speedup, but is very general and can be applied to several different problems. Since then many quantum algorithms have been developed, some of them even with exponential speed-up. Some nice examples include quantum algorithms for Pell's equations [9], for Boolean formula evaluation [6, 14], for solving linear systems of equations [10], for subset finding [2], for finding triangles in graphs [1, 11], etc. Similarly to classical computing, where different approaches can be employed for designing efficient algorithms, also in quantum computing there are different approaches for designing quantum algorithms. The "Quantum Algorithm Zoo" is a website that keeps track of all quantum algorithms,[2] classified according to the employed technique. It is interesting to notice that Shor's and Grover's algorithms are still good representatives of the main techniques for quantum algorithm design, the former representing the category of algebraic and number theoretic algorithms, and the later representing the category of oracular algorithms. Therefore, every student of quantum computing should study these two algorithms.

Grover's algorithm has been physically implemented for the first time in 1998 using nuclear magnetic resonance [3], and Shor's algorithm have been implemented in 2001 also using this same technique [16]. In both cases, only small instances of the problems were addressed. Apart from that, there were small implementations of quantum gates and protocols over few qubits. Because of that, in the beginning, quantum computing was considered a technological promise for the distant future.

---

[2]S. Jordan. "Quantum Algorithm Zoo", https://math.nist.gov/quantum/zoo/.

Even researchers in the field believed that practical implementations of quantum computers would still take at least several decades to come. Recently, however, important milestones in the construction of quantum hardware have been achieved, so the quantum revolution seems to be much closer now. Intel announced a 49-qubit computer[3]; IBM released quantum computers up to 20 qubits for public use in the cloud, and announced the tests of a 50-qubit computer[4]; researchers from Google also announced that they are testing a 72-qubit computer[5]! It is important to notice that the largest simulation of quantum computers is of a 56-qubit system, performed by IBM using a supercomputer.[6] Therefore, if Google keeps the promise of delivering a 72-qubit quantum computer, there will be no classical computer capable of simulating it, and therefore the so-called quantum supremacy will be achieved. Rigetti is planning to build and deploy a 128-qubit quantum processing unit in the near future.[7]

There are still many challenges to overcome in the journey towards a general-purpose quantum computer. The number of qubits increased considerably in the last few years, but the noise rates achieved by the quantum logical gates are still very high, and the decoherence times are still too short for most problems. However, there is no doubt that practical quantum computing seems much closer to reality nowadays. Some large companies are throwing money on quantum computing. New companies are emerging in this field. There is even a conference for quantum computing in business[8]!

Despite the increasing importance of quantum computing, students usually face great challenges in the beginning of their training in this field. Quantum mechanics is usually recognized by the general public as one of the hardest topics of modern science, and it sometimes frightens students of computer science or mathematics. However, a deep knowledge of quantum mechanics is not a requirement to start in quantum computing. It is important to have good mathematical skills, specially in linear algebra, and to learn some basic rules of quantum mechanics. Similarly, some concepts of computer science can be hard for students of physics or engineering to grasp. However, a deep knowledge of complexity classes, Turing machines, and computer programming is not a requirement for beginner students of quantum computing.

---

[3]Intel Newsroom, "The Future of Quantum Computing is Counted in Qubits", https://intel.ly/2JNHeqK.

[4]E. Conover, "Quantum computers take a step forward with a 50-qubit prototype", https://bit.ly/2yOcBk2.

[5]E. Conover, "Google moves toward quantum supremacy with 72-qubit computer", https://bit.ly/2FiFp7q.

[6]C. Q. Choi, "IBM Simulates a 56-Qubit Machine", https://bit.ly/2yYyL2i.

[7]E. Newcomer, "Quantum Computers Today Aren't Very Useful. That Could Change". https://bloom.bg/2M2gvws.

[8]The first conference of Quantum Computing for Business took place in California in 2017, with guest speakers from Airbus, Goldman Sachs, Volkswagen, among others. More information in https://www.q2b.us/.

It is important to be prepared for this new technology and this book covers all the essential topics to get started. It is organized as follows. In Chap. 2, we introduce the basic notions about quantum mechanics, quantum logic gates, and quantum circuits that will be necessary for the rest of the book. In Chap. 3, we present Grover's algorithm for searching unordered lists. In Chap. 4, the quantum Fourier transform and Shor's algorithm for integer factorization are given. In Chap. 5, we present quantum walks, the quantum counterpart of classical random walks. Even though the present book is a primer, therefore focused on the fundamentals of the field, in this chapter we decided to give more attention to the staggered model of quantum walks, since it generalizes the most important cases of all the other quantum walk models, and thus the beginner can benefit from this approach. Finally, Chap. 6 presents our final discussions and suggests further topics of study to the reader.

References

1. Carette, T., Laurière, M., Magniez, F.: Extended learning graphs for triangle finding. In: 34th Symposium on Theoretical Aspects of Computer Science, STACS 2017, March 8–11, 2017, Hannover, pp. 20:1–20:14 (2017). https://doi.org/10.4230/LIPIcs.STACS.2017.20
2. Childs, A.M., Eisenberg, J.M.: Quantum algorithms for subset finding. Quantum Inf. Comput. **5**(7), 593–604 (2005)
3. Chuang, I.L., Gershenfeld, N., Kubinec, M.: Experimental implementation of fast quantum searching. Phys. Rev. Lett. **80**, 3408–3411 (1998). https://doi.org/10.1103/PhysRevLett.80.3408
4. Deutsch, D.: Quantum theory, the Church-Turing principle and the universal quantum computer. Proc. R. Soc. A Math. Phys. Eng. Sci. **400**(1818), 97–117 (1985). https://doi.org/10.1098/rspa.1985.0070
5. Deutsch, D., Jozsa, R.: Rapid solution of problems by quantum computation. Proc. R. Soc. Lond. A Math. Phys. Eng. Sci. **439**(1907), 553–558 (1992). https://doi.org/10.1098/rspa.1992.0167
6. Farhi, E., Goldstone, J., Gutmann, S.: A quantum algorithm for the hamiltonian nand tree. Technical Report, Massachusetts Institute of Technology (2007). MIT-CTP/3813
7. Feynman, R.P.: Simulating physics with computers. Int. J. Theor. Phys. **21**(6), 467–488 (1982). https://doi.org/10.1007/BF02650179
8. Grover, L.K.: Quantum mechanics helps in searching for a needle in a haystack. Phys. Rev. Lett. **79**(2), 325–328 (1997). https://doi.org/10.1103/PhysRevLett.79.325
9. Hallgren, S.: Polynomial-time quantum algorithms for Pell's equation and the principal ideal problem. In: Proceedings of the Thiry-fourth Annual ACM Symposium on Theory of Computing, STOC '02, pp. 653–658. ACM, New York (2002). https://doi.org/10.1145/509907.510001
10. Harrow, A.W., Hassidim, A., Lloyd, S.: Quantum algorithm for linear systems of equations. Phys. Rev. Lett. **103**, 150,502 (2009). https://doi.org/10.1103/PhysRevLett.103.150502
11. Magniez, F., Santha, M., Szegedy, M.: Quantum algorithms for the triangle problem. In: Proceedings of the Sixteenth Annual ACM-SIAM Symposium on Discrete Algorithms, SODA '05, pp. 1109–1117. Society for Industrial and Applied Mathematics, Philadelphia (2005)
12. Manin, Y.: Computable and Uncomputable (in Russian). Sovetskoye Radio, Moscow (1980)
13. Moore, G.E.: Cramming more components onto integrated circuits. Electronics **38**(8), 114–117 (1965)

14. Reichardt, B.W.: Reflections for quantum query algorithms. In: Proceedings of the Twenty-second Annual ACM-SIAM Symposium on Discrete Algorithms, SODA '11, pp. 560–569. Society for Industrial and Applied Mathematics, Philadelphia (2011)
15. Shor, P.: Algorithms for quantum computation: discrete logarithms and factoring. In: Proceedings 35th Annual Symposium on Foundations of Computer Science, pp. 124–134. IEEE Computer Society, Los Alamitos (1994). https://doi.org/10.1109/SFCS.1994.365700
16. Vandersypen, L.M.K., Steffen, M., Breyta, G., Yannoni, C.S., Sherwood, M.H., Chuang, I.L.: Experimental realization of Shor's quantum factoring algorithm using nuclear magnetic resonance. Nature **414**(6866), 883–887 (2001). https://doi.org/10.1038/414883a

Chapter 2
Bits and Qubits

Quantum computing is an interdisciplinary field of research, which lies in the intersection of computer science and physics. Therefore, it is natural that most people beginning in this area should feel a need of learning some fundamental concepts of either computer science or quantum mechanics before diving into the matter of quantum computing. In order to keep this book as self-contained as possible, we briefly review those concepts in this chapter. We start with a review of classical circuits and reversible computing, in Sect. 2.1. All the concepts of quantum mechanics necessary for understanding the book are summarized and explained in Sect. 2.2. In the rest of this chapter, we discuss quantum gates and the basic tools for studying quantum computing.

2.1 The Classical Computer and Reversibility

A classical computer can be understood in a very broad sense as a machine that reads a certain amount of data encoded as zeroes and ones, performs calculations, and then outputs data as zeroes and ones again. Zeroes and ones are states of some physical quantity, such as the electric potential in classical computers. Internally, a zero may be a state of low electric potential while a one may be a state of high electric potential, for example. The actual physical interpretation depends on the technology used to build the hardware, and is usually abstracted. The point is that *some* classical physical system must be used to represent information, and understanding this is crucial to the generalization we will discuss ahead.

Of course, zeroes and ones form a binary number which can be converted to decimal notation. Thus, we may think of the computer as calculating a function

$$f : \{0, \ldots, N-1\} \to \{0, \ldots, N-1\}, \tag{2.1}$$

© The Author(s), under exclusive license to Springer Nature Switzerland AG 2019
F. de Lima Marquezino et al., *A Primer on Quantum Computing*, SpringerBriefs in Computer Science, https://doi.org/10.1007/978-3-030-19066-8_2

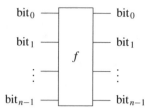

Fig. 2.1 Outline of a classical computer. Since we are more interested on the *reversible* model of classical computing, here we assume that the input and the output of the computer have the same number of bits

where N is a number of the form 2^n, and n is the number of bits in the computer memory. In this description, f must be a function because the computer cannot generate two or more different outputs from the same input. We assume without loss of generality that the domain and codomain of f are of the same size. In other words, we assume that both the input and the output of the computer have the same number of bits. This assumption will be important later on, since we will be more interested on the *reversible* model of classical computing.

The calculation process we just described is depicted in Fig. 2.1, where on the left hand side we have the value of each bit (zero or one). The process of calculation goes from left to right, and the output is stored in the same bits on the right hand side.

Usually f is given in terms of elementary blocks—or *logic gates*—that can be implemented in practice by using transistors and other electrical devices. The blocks are the AND, OR and NOT gates, known as universal gates. This set could be reduced further, since OR can be written in terms of AND and NOT. For example, a circuit to add two one-bit numbers modulo two is given in Fig. 2.2. The possible inputs are 00, 01, 10, 11, and the corresponding outputs are 00, 01, 11, 10. The inputs may be prepared, for instance, by creating electric potential gaps, which create electric currents that propagate through the wires towards right. The gates are activated as time goes by. The meter symbols on the right indicate that measurements of the electric potential are performed, which tell whether the output value of each bit is zero or one. The second bit gives the result of the calculation. The wire for the output of the first bit is artificial and unnecessary; at this point, it is there simply to have the codomain size of the function f equal to the domain size. This circuit, without the first bit output wire, is the circuit for the XOR (exclusive OR) gate in terms of the universal gates.

The circuit of Fig. 2.2 is irreversible, since the gates AND and OR are irreversible. If the output of the AND gate is zero, nobody can tell what was the input, and similarly when the output of the OR gate is one. This means that the physical theory which describes the processes in Fig. 2.2 must be irreversible. Then, the AND and OR gates cannot be straightforwardly generalized to quantum gates, which must be reversible ones.

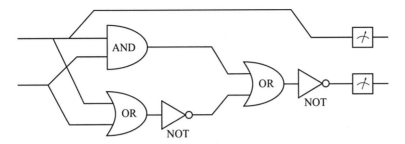

Fig. 2.2 A circuit to add two one-bit numbers modulo two

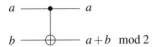

Fig. 2.3 Classical controlled-NOT (CNOT) gate denoting a positive control on the upper line, and target on the bottom line

However, the circuit of Fig. 2.2 can be made reversible. Although the following description is unnecessary from the classical point of view, it helps the quantum generalization performed in the next sections. We employ the controlled-NOT (CNOT) gate of Fig. 2.3. The bits a and b assume values either zero or one. The value of the first bit—called the control bit—never changes in this gate, while the value of the second bit—called the target bit—is flipped only if $a = 1$. If $a = 0$, then nothing happens to both bits. This gate, denoted by $\oplus$, is a NOT gate controlled by the value of the first bit.[1] Now it is easy to verify that the value of the second bit for this gate is $a + b \mod 2$. The CNOT gate is not a universal building block for classical circuits, but its quantum counterpart is a basic block of quantum circuits.

We have described the reversible counterpart of the XOR gate. What is the reversible counterpart of the AND gate? The answer employs the *Toffoli* gate which is a generalization of the CNOT gate with two control bits instead of one. The Toffoli gate is depicted in Fig. 2.4. The value of the third bit—the target—is inverted only if both a and b are equal to one, otherwise it does not change. The following table describes all possible inputs and the corresponding outputs, where the most significant bit corresponds to the upper line of the circuit:

$$000 \rightarrow 000$$

$$001 \rightarrow 001$$

$$010 \rightarrow 010$$

---

[1]Sometimes it may be convenient to have a CNOT with negative control, so that the target bit is flipped if and only if the control bit is zero. A negative control is usually represented in the circuit model by replacing the full bullet • by an empty bullet ○ on the control line.

Fig. 2.4 Classical Toffoli gate denoting positive controls on the upper and middle lines, and target on the bottom line

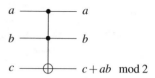

$$011 \to 011$$
$$100 \to 100$$
$$101 \to 101$$
$$110 \to 111$$
$$111 \to 110$$

The AND gate can be replaced by the Toffoli gate simply by taking $c = 0$. The output of the third bit is then a AND b. The reversible circuit for the OR gate is a little cumbersome because it requires more than one Toffoli gate, so we will not describe it here. For a reference, we suggest [38].

Another feature implicit in Fig. 2.2 that cannot be performed in quantum circuits is bifurcation of wires—known as FANOUT. This is equivalent to copying the value of a bit, and there is no problem to do this classically. However, this is forbidden in quantum circuits, due to an important result of quantum mechanics known as the *no-cloning theorem*. Notice that a classical FANOUT can be obtained from the CNOT gate by taking $b = 0$.

Consider again Fig. 2.1. If the computer has n bits, there are 2^n possible inputs. For each input there are 2^n possible outputs, therefore the number of possible functions f that can be calculated is 2^{n2^n}. All these functions can be reduced to circuits using the universal gates. That is what a reversible classical computer with n bits can do: calculate 2^{n2^n} functions! This number is astronomical for computers with a few gigabytes, that is a typical memory size for good personal computers nowadays.

Another important issue is how fast the computer can calculate these functions. If we assume that each elementary gate takes constant time, then the answer depends on the number of gates used in the circuit for f. If the number of elementary gates increases polynomially with n, we say that the circuit is "efficient". If the number of gates increases exponentially with n, the circuit is "inefficient". This is a very coarse method to measure the efficiency, but it is useful for theoretical analysis when n is large. Note that we are thinking of computers of variable size, which is not the case in practice. In fact, instead of referring to actual computers, it is better to use a Turing machine, which is an abstract model for computers and software as a whole. Similarly, quantum computers and their software are abstractly modeled as the quantum Turing machine. The classical theory of complexity classes and its quantum counterpart address this kind of problems.

For a description of the classical Turing machine see Papadimitriou [37], Lewis and Papadimitriou [26], or Arora and Barak [4]. The quantum Turing machine was first defined by Deutsch [14], and the reader may also refer to Miszczak [32], Bernstein and Vazirani [7], or Vazirani [45] for details.

All calculations that can be performed in a classical computer can also be performed in a quantum computer. One simply replaces the irreversible gates of the classical computer with their reversible counterparts. The new circuit can be implemented in a quantum computer. But there is no advantage in this procedure: why build a very expensive quantum machine which behaves classically? The appeal of quantum computation is the possibility of quantum algorithms be faster than classical ones. The quantum algorithms must use quantum features not available in classical computers, such as quantum parallelism and entanglement, in order to enhance the computation. On the other hand, a naïve use of quantum features does not guarantee any improvements.

Historically, there are two remarkable classes of successful quantum algorithms: the database search algorithms and the algorithms for finding the generators of a normal subgroup of a given group. Shor's algorithms for integer factorization and discrete logarithm are special cases of this latter class [43]. More recently, other techniques have been developed. Quantum walks, the quantum counterpart of random walks, have been successfully used to design quantum algorithms for search on graphs [29, 39], element distinctness [3], triangle finding [28], and many other problems [33]. Adiabatic quantum computing, analogous to the classical technique of simulated annealing, has been applied for discrete optimization [9, 41]. The quantum algorithm zoo[2] is growing fast.

Exercise 2.1 Build the classical circuit for the OR operation in terms of reversible gates. *Hint:* Since the OR operation is irreversible, you will need to use an ancillary bit.

Exercise 2.2 Rewrite the circuit of Fig. 2.2 as a reversible circuit.

2.2   Review of Quantum Mechanics for Quantum Computation

At first sight, quantum computing may seem intimidating for a non-physicist. After all, quantum mechanics has the reputation of being an extremely hard topic to grasp. Although it can be true in general, we must keep in mind that in order to start studying quantum computing, we only need a subset of the topics that are usually studied in quantum mechanics. The game of chess is a good analogy: we do not need to know all the strategies of the game in order to play the game, although we definitely need to know the game *rules*! The rules of the game for quantum

---

[2]Currently, there is an excellent website known as the *Quantum Algorithm Zoo*, available at http:// math.nist.gov/quantum/zoo/.

mechanics can be summarized in at least four *postulates*. In this section, we give
a gentle introduction to the essential ideas of quantum mechanics that will be
necessary to understand the rest of the book. We also introduce the proper notation
and terminology. There are many good references for quantum mechanics, such as
Cohen-Tannoudji et al. [12] and Messiah [30]. There are also several books on linear
algebra, for instance Lang [25] and Hoffman and Kunze [20]. Aaronson's book [1]
provides an interesting and modern approach covering all those topics with focus
on quantum computing.

It is convenient to start our journey through quantum mechanics by establishing
an analogy with classical computing. In classical computers, a bit can assume only
values 0 or 1. In quantum computers, the values 0 and 1 are replaced by the vectors
$|0\rangle$ and $|1\rangle$. This notation for vectors is called the Dirac notation and is standard
in quantum mechanics. The term bit is replaced by *qubit*, short of *quantum bit*.
Differently from the classical bit, the quantum bit $|\psi\rangle$ can also be in a linear
combination of the vectors $|0\rangle$ and $|1\rangle$, that is, we can write

$$|\psi\rangle = \alpha|0\rangle + \beta|1\rangle, \tag{2.2}$$

where α and β are complex numbers. The vector $|\psi\rangle$ is said to be a *superposition*
of the vectors $|0\rangle$ and $|1\rangle$ with *amplitudes* α and β. Thus, $|\psi\rangle$ is a vector in a two-
dimensional complex vector space, where $\{|0\rangle, |1\rangle\}$ forms an orthonormal basis,
called the *computational basis*. Notice that the state $|0\rangle$ is not the zero vector, but
simply the first vector of the basis. The matrix representations of the vectors $|0\rangle$
and $|1\rangle$ are usually given by

$$|0\rangle = \begin{bmatrix} 1 \\ 0 \end{bmatrix} \quad \text{and} \quad |1\rangle = \begin{bmatrix} 0 \\ 1 \end{bmatrix}. \tag{2.3}$$

In quantum mechanics, vectors are systematically called *states*, so we shall use this
term from now on. At this point, we can enunciate the first postulate of quantum
mechanics.

Postulate (State-Space) An isolated physical system is associated to a Hilbert
space, called *state space*. The state of the system, called *state vector*, is fully
described by a unit vector in the state space.

A complex vector space V is a Hilbert space if there is an *inner product*, written
in the form $\langle\varphi|\psi\rangle$, defined by the following rules:

1. $\langle\psi|\varphi\rangle = \langle\varphi|\psi\rangle^*$,
2. $\langle\varphi|(\alpha|u\rangle + \beta|v\rangle)) = \alpha\langle\varphi|u\rangle + \beta\langle\varphi|v\rangle$,
3. $\langle\varphi|\varphi\rangle > 0$ if $|\varphi\rangle \neq 0$,

where $\alpha, \beta \in \mathbb{C}$, $|\varphi\rangle, |\psi\rangle, |u\rangle, |v\rangle \in V$, and $*$ is the complex conjugate operation.

The *norm* of a vector $|\varphi\rangle$ is given by

$$\||\varphi\rangle\| = \sqrt{\langle\varphi|\varphi\rangle}. \tag{2.4}$$

The notation $\langle\varphi|$ is used for the *dual vector* to the vector $|\varphi\rangle$. The dual is a linear operator from the vector space V to the complex numbers satisfying

$$\langle\varphi|(|v\rangle) = \langle\varphi|v\rangle, \quad \forall|v\rangle \in V. \tag{2.5}$$

Given two vectors $|\varphi\rangle$ and $|\psi\rangle$ in a vector space V, there is also an *outer product* $|\psi\rangle\langle\varphi|$, defined as a linear operator on V satisfying

$$(|\psi\rangle\langle\varphi|)|v\rangle = |\psi\rangle\langle\varphi|v\rangle, \quad \forall|v\rangle \in V. \tag{2.6}$$

If $|\varphi\rangle = \alpha|0\rangle + \beta|1\rangle$ and $|\psi\rangle = \gamma|0\rangle + \delta|1\rangle$, then the matrix representations for inner and outer products are, respectively,

$$\langle\varphi|\psi\rangle = \begin{bmatrix} \alpha^* & \beta^* \end{bmatrix}\begin{bmatrix} \gamma \\ \delta \end{bmatrix} = \alpha^*\gamma + \beta^*\delta, \tag{2.7}$$

and

$$|\varphi\rangle\langle\psi| = \begin{bmatrix} \alpha \\ \beta \end{bmatrix}\begin{bmatrix} \gamma^* & \delta^* \end{bmatrix} = \begin{bmatrix} \alpha\gamma^* & \alpha\delta^* \\ \beta\gamma^* & \beta\delta^* \end{bmatrix}. \tag{2.8}$$

The matrix of the outer product is obtained by usual matrix multiplication of a column matrix by a row matrix. In this case, however, we can replace the matrix multiplication by the tensor product, i.e., $|\varphi\rangle\langle\psi| = |\varphi\rangle \otimes \langle\psi|$, which shall be explained later on. Notice the complex conjugation in the process of taking the dual.

Looking back to our previous example, we notice that a qubit is represented by a vector on a two-dimensional complex Hilbert space. That would be the case of electron spin, for instance. However, the postulate allows for much more complicated physical systems. For the purposes of this book, it will be enough to consider only finite Hilbert spaces. However, for some applications, the dimension of the Hilbert space required to represent a given physical system may be infinite.

It is also interesting to notice that we mentioned an *isolated* physical system when stating the postulate. By isolated (or *closed*) system we understand a physical system which neither influences nor is influenced by the outside. In practice, the system is never completely isolated, therefore the state-space postulate is an idealization.

Back to our two-dimensional example, we have in Fig. 2.5 a visualization for a qubit in the case where the amplitudes are real numbers. It is not a coincidence that the state $|\psi\rangle$ is being represented as a unit vector. The reason is that the amplitudes of a quantum state are associated to certain probabilities, and thus they must respect some constraints. In order to understand this, we should first think about the physical interpretation of the superposition. We can say that $|\psi\rangle$ coexists in two states: $|0\rangle$ and $|1\rangle$. It is similar to a coin that is partially heads up and partially tails up simultaneously. We cannot push further the analogy simply because quantum phenomena do not have a classical analogue in general. The state $|\psi\rangle$ can store a huge quantity of information in its coefficients α and β, but this information lives in the quantum level, which is microscopic—usually quantum effects appear in atomic dimensions. To bring quantum information to the classical level, one must measure the qubit. Quantum mechanics tells us that the measurement process inevitably disturbs a qubit state, producing a non-deterministic collapse of $|\psi\rangle$ to either $|0\rangle$ or $|1\rangle$. Roughly, we can say that one gets $|0\rangle$ with probability $|\alpha|^2$ or $|1\rangle$ with probability $|\beta|^2$, considering a measurement in the computational basis. At this point, we can enunciate another postulate of quantum mechanics.

Postulate (Measurement) A *projective measurement* is described by an Hermitean operator $\mathcal{O}$ acting over the state space of the system being measured. This operator $\mathcal{O}$ is called *observable* and has a diagonal representation

$$\mathcal{O} = \sum_{\lambda} \lambda P_{\lambda}, \tag{2.9}$$

where P_{λ} is the projector on the eigenspace of $\mathcal{O}$ associated with the eigenvalue λ. The possible results of measurement of the observable $\mathcal{O}$ are the eigenvalues λ. If immediately before the measurement the system is in state $|\psi\rangle$, then the probability of obtaining result λ is given by $\| P_{\lambda} |\psi\rangle \|^2$, which is equivalent to

$$p_{\lambda} = \langle \psi | P_{\lambda} | \psi \rangle. \tag{2.10}$$

If the result of measurement is λ, then the state of the system is irreversibly collapsed to

$$\frac{1}{\sqrt{p_{\lambda}}} P_{\lambda} |\psi\rangle . \tag{2.11}$$

When the context is clear, it is not unusual to refer to the result of the measurement as the collapsed state. However, the reader should keep in mind that the result is rigorously described as a real number, namely the eigenvalue of an Hermitean operator.

In the previous example, the measurement of one qubit in the computational basis is represented by an observable

$$Z = \begin{pmatrix} 1 & 0 \\ 0 & -1 \end{pmatrix}. \tag{2.12}$$

Notice that Z is an Hermitean matrix, with eigenvalues $+1$ and -1 associated to eigenvectors $|0\rangle$ and $|1\rangle$, respectively. Thus, the projectors on the eigenspaces of Z are $P_{+1} = |0\rangle \langle 0|$ and $P_{-1} = |1\rangle \langle 1|$.

Recall that the non-deterministic collapse does not allow one to determine the values of α and β before the measurement. They are inaccessible via measurements unless one has many copies of the same state. Two successive measurements of the

Fig. 2.5 Visualization of a
qubit in the computational
basis for the case where α, β
are real numbers

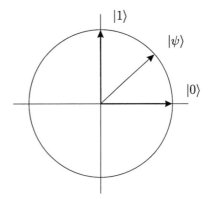

same qubit give the same output. If $|\alpha|^2$ and $|\beta|^2$ are probabilities and there are only two possible outputs, then

$$|\alpha|^2 + |\beta|^2 = 1. \tag{2.13}$$

Calculating the norm of $|\psi\rangle$, the previous equation gives

$$\| \, |\psi\rangle \, \| = \sqrt{|\alpha|^2 + |\beta|^2}$$
$$= 1. \tag{2.14}$$

In the general case, where α and β are complex numbers, there is still a geometrical representation for a qubit. In order to understand this representation, let us first rewrite the amplitudes as

$$\alpha = |\alpha|e^{i\gamma},$$
$$\beta = |\beta|e^{i(\gamma+\varphi)}, \tag{2.15}$$

and let us define

$$\cos \frac{\theta}{2} = |\alpha|$$
$$\sin \frac{\theta}{2} = |\beta|. \tag{2.16}$$

Using these definitions, it is possible to prove [36] that Eq. (2.2) can be rewritten as

$$|\psi\rangle = e^{i\gamma} \left(\cos \frac{\theta}{2} |0\rangle + e^{i\varphi} \sin \frac{\theta}{2} |1\rangle \right). \tag{2.17}$$

Fig. 2.6 The Bloch sphere representing the qubit from Eq. (2.17)

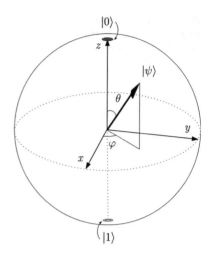

For purposes of visualization, the factor $e^{i\gamma}$ outside parenthesis—called *global phase factor*—can be ignored, because it has no observable effect on the state according to the postulate of measurement. Therefore, we can represent any qubit on a sphere, called the Bloch sphere, as shown in Fig. 2.6.

A measurement is not the only way that one can interact with a qubit. If one does not obtain any information about the state of the qubit, the interaction changes the values of α and β keeping the constraint of Eq. (2.13). The most general transformation of this kind is a linear transformation U that takes unit vectors into unit vectors. Such a transformation is called *unitary* and satisfies

$$U^{\dagger}U = UU^{\dagger} = I, \tag{2.18}$$

where $U^{\dagger}$ is the adjoint transformation of U. One can always find a matrix representation for any linear transformation. In this book, unless stated otherwise, we use the same symbol when referring to the linear transformation or to its corresponding matrix representation. The matrix representation of $U^{\dagger}$ corresponds to the conjugate transpose of the matrix representation of U. The next postulate of quantum mechanics deals with the evolution of closed quantum systems.

Postulate (Evolution I) The time evolution of an isolated quantum system is described by Schrödinger's equation,

$$H \left| \psi(t) \right\rangle = i\hbar \frac{d}{dt} \left| \psi(t) \right\rangle, \tag{2.19}$$

where H is an Hermitean operator called Hamiltonian, and $\hbar$ equals Plank's constant divided by 2π.

For many applications in the area of quantum algorithms, we may assume that the Hamiltonian is time-independent. In that case, the solution to Schrödinger's equation is given by

$$|\psi(t)\rangle = \exp(i\hbar Ht)|\psi(0)\rangle,$$
$$= U|\psi(0)\rangle, \tag{2.20}$$

where U is a unitary operator. The above equation leads us to an alternative formulation of the evolution postulate, as follows.

Postulate (Evolution II) The time evolution of an isolated quantum system is described by a unitary quantum operator. If the state of the system at time t_0 is described by $|\psi_0\rangle$, and the state of the system at time t_1 is described by $|\psi_1\rangle$, then the evolution of the system depends only on t_0 and t_1, and is described by a unitary operator U, such that

$$|\psi_1\rangle = U|\psi_0\rangle. \tag{2.21}$$

An algorithm is a finite sequence of interactions that must be performed on an input state in order to change it into a output state that represents the solution for a certain problem. Therefore, a quantum algorithm is ultimately a sequence of unitary operators.

So far we are dealing with one-qubit quantum computers. To consider the multiple qubit case, it is necessary to introduce the concept of *tensor product* or *Kronecker product*. Suppose V and W are complex vector spaces of dimensions m and n, respectively. The tensor product $V \otimes W$ is an mn-dimensional vector space. The elements of $V \otimes W$ are linear combinations of tensor products $|v\rangle \otimes |w\rangle$, satisfying the following properties:

1. $z(|v\rangle \otimes |w\rangle) = (z|v\rangle) \otimes |w\rangle = |v\rangle \otimes (z|w\rangle)$,
2. $(|v_1\rangle + |v_2\rangle) \otimes |w\rangle = (|v_1\rangle \otimes |w\rangle) + (|v_2\rangle \otimes |w\rangle)$,
3. $|v\rangle \otimes (|w_1\rangle + |w_2\rangle) = (|v\rangle \otimes |w_1\rangle) + (|v\rangle \otimes |w_2\rangle)$,

where $z \in \mathbb{C}$, and $|v\rangle, |v_1\rangle, |v_2\rangle \in V$, and $|w\rangle, |w_1\rangle, |w_2\rangle \in W$. We also use the notations $|v\rangle|w\rangle$, or $|v, w\rangle$, or $|vw\rangle$ for the tensor product $|v\rangle \otimes |w\rangle$. Note that the tensor product is non-commutative, so the notation must preserve the ordering.

Given two linear operators A and B defined on the vector spaces V and W, respectively, we can define the linear operator $A \otimes B$ over the vector space $V \otimes W$ as

$$(A \otimes B)(|v\rangle \otimes |w\rangle) = A|v\rangle \otimes B|w\rangle, \tag{2.22}$$

where $|v\rangle \in V$ and $|w\rangle \in W$. The matrix representation of $A \otimes B$ is given by

$$A \otimes B = \begin{bmatrix} A_{11}B & \cdots & A_{1m}B \\ \vdots & \ddots & \vdots \\ A_{m1}B & \cdots & A_{mm}B \end{bmatrix}, \tag{2.23}$$

where A is an $m \times m$ matrix and B is a $n \times n$ matrix. Hence, matrix $A \otimes B$ has dimension $mn \times mn$. For example, given

$$A = \begin{bmatrix} 0 & 1 \\ 1 & 0 \end{bmatrix} \quad \text{and} \quad B = \begin{bmatrix} 1 & 0 & 0 \\ 0 & 1 & 0 \\ 0 & 0 & 1 \end{bmatrix},$$

the tensor product $A \otimes B$ is

$$A \otimes B = \begin{bmatrix} 0 & 1 \\ 1 & 0 \end{bmatrix} \otimes \begin{bmatrix} 1 & 0 & 0 \\ 0 & 1 & 0 \\ 0 & 0 & 1 \end{bmatrix} = \begin{bmatrix} 0 & 0 & 0 & 1 & 0 & 0 \\ 0 & 0 & 0 & 0 & 1 & 0 \\ 0 & 0 & 0 & 0 & 0 & 1 \\ 1 & 0 & 0 & 0 & 0 & 0 \\ 0 & 1 & 0 & 0 & 0 & 0 \\ 0 & 0 & 1 & 0 & 0 & 0 \end{bmatrix}. \tag{2.24}$$

The formula of Eq. (2.23) can also be used for non-square matrices, such as the tensor product of two vectors. For example, if we have a two-qubit quantum computer and the first qubit is in the state $|0\rangle$ and the second is in the state $|1\rangle$, then the quantum computer is in the state $|0\rangle \otimes |1\rangle$, given by

$$|0\rangle \otimes |1\rangle = \begin{bmatrix} 1 \\ 0 \end{bmatrix} \otimes \begin{bmatrix} 0 \\ 1 \end{bmatrix} = \begin{bmatrix} 0 \\ 1 \\ 0 \\ 0 \end{bmatrix}. \tag{2.25}$$

The resulting vector is in a four-dimensional vector space.

At this point, we can finally state the last postulate that we will need for our studies.

Postulate (Composition) The state space of a *composite system* is given by the tensor product of the state space of its components. If $|\psi_0\rangle, |\psi_1\rangle, \ldots, |\psi_{n-1}\rangle$ are the states of n isolated systems, then $|\psi_0\rangle \otimes |\psi_1\rangle \otimes \ldots \otimes |\psi_{n-1}\rangle$ is the state of the composite system.

The general state $|\psi\rangle$ of a two-qubit quantum computer is a superposition of the states $|00\rangle, |01\rangle, |10\rangle,$ and $|11\rangle$, namely

$$|\psi\rangle = \alpha|00\rangle + \beta|01\rangle + \gamma|10\rangle + \delta|11\rangle, \tag{2.26}$$

with the constraint

$$|\alpha|^2 + |\beta|^2 + |\gamma|^2 + |\delta|^2 = 1.$$

Regarding the zeroes and ones as constituting the binary expansion of an integer, we can replace the representations of states

$$|00\rangle, \ |01\rangle, \ |10\rangle, \ |11\rangle, \tag{2.27}$$

by the shorter forms

$$|0\rangle, \ |1\rangle, \ |2\rangle, \ |3\rangle, \tag{2.28}$$

in decimal notation, which is handy in some formulas.

In general, the state $|\psi\rangle$ of an n-qubit quantum computer is a superposition of the 2^n states $|0\rangle, |1\rangle, \ldots, |2^n - 1\rangle$, namely

$$|\psi\rangle = \sum_{i=0}^{2^n-1} \alpha_i |i\rangle, \tag{2.29}$$

with amplitudes α_i constrained to

$$\sum_{i=0}^{2^n-1} |\alpha_i|^2 = 1. \tag{2.30}$$

Recall that the orthonormal basis $\{|0\rangle, \ldots, |2^n - 1\rangle\}$ is the computational basis in decimal notation. The state of an n-qubit quantum computer is a vector in a 2^n-dimensional complex vector space. When the number of qubits increases linearly, the dimension of the associated vector space increases exponentially. When dealing with multiple qubits, it is sometimes useful to define the notation $|\psi\rangle^{\otimes n}$ meaning $|\psi\rangle \otimes |\psi\rangle \otimes \ldots \otimes |\psi\rangle$, repeated n times, and analogously for linear operator, such as $U^{\otimes n}$.

As before, a projective measurement of a generic state $|\psi\rangle$ on the computational basis yields the result $|i_0\rangle$ with probability $|\alpha_{i_0}|^2$, where $0 \leq i_0 < 2^n$. Usually, the measurement is performed qubit by qubit yielding zeroes or ones that are read together to form i_0. We stress again a very important feature of the measurement process. The amplitudes of the state $|\psi\rangle$ as it is before measurement are inaccessible. The measurement process inevitably disturbs $|\psi\rangle$ forcing it to collapse to one vector of the computational basis. This collapse is non-deterministic, with the probabilities given by the squared norms of the corresponding amplitudes in $|\psi\rangle$.

When we consider quantum computers with at least two qubits, an interesting phenomenon may occur. Consider that we have a two-qubit quantum computer, with the first qubit in the state

$$|\varphi\rangle = a|0\rangle + b|1\rangle \tag{2.31}$$

and the second qubit in the state

$$|\psi\rangle = c|0\rangle + d|1\rangle. \tag{2.32}$$

Thus, the state of the quantum computer is the tensor product

$$|\varphi\rangle \otimes |\psi\rangle = (a|0\rangle + b|1\rangle) \otimes (c|0\rangle + d|1\rangle)$$
$$= ac|00\rangle + ad|01\rangle + bc|10\rangle + bd|11\rangle. \tag{2.33}$$

Note that a general two-qubit state as in Eq. (2.26) is of the form of Eq. (2.33) if and only if

$$\alpha = ac,$$
$$\beta = ad,$$
$$\gamma = bc,$$
$$\delta = bd. \tag{2.34}$$

From these identities, we conclude that a general two-qubit state as in Eq. (2.26) is of the form of Eq. (2.33) if and only if

$$\alpha\delta = \beta\gamma. \tag{2.35}$$

Therefore, the general two-qubit state is not necessarily a product of two one-qubit states. For example, the state

$$|\beta_{00}\rangle = \frac{|00\rangle + |11\rangle}{\sqrt{2}} \tag{2.36}$$

is a valid state for a two-qubit quantum computer, even though it cannot be written as a product of two qubits in the form of Eq. (2.33). Such non-product states of two or more qubits are called *entangled* states. In fact, the above state is very special, since it belongs to the special class of entangled states known as the *Bell states* [36]. The other Bell states are

$$|\beta_{01}\rangle = \frac{|01\rangle + |10\rangle}{\sqrt{2}}, \tag{2.37}$$

$$|\beta_{10}\rangle = \frac{|00\rangle - |11\rangle}{\sqrt{2}}, \tag{2.38}$$

$$|\beta_{11}\rangle = \frac{|01\rangle - |10\rangle}{\sqrt{2}}. \tag{2.39}$$

The entangled states play an essential role in quantum computation. Quantum computers that do not use entanglement cannot be exponentially faster than classical computers. On the other hand, a naive use of entanglement does not guarantee any improvements.

After the above review of quantum mechanics, we are ready to outline the quantum computer. In Fig. 2.7, we are taking a non-entangled input, which is quite reasonable. In fact, $|\psi_i\rangle$ is either $|0\rangle$ or $|1\rangle$ generally. State $|\psi\rangle$, on the right hand

Fig. 2.7 The sketch of the quantum computer. We consider the input non-entangled, which is reasonable in general. On the other hand, the output is entangled in general. The measurement of the state $|\psi\rangle$, not shown here, returns zeroes and ones

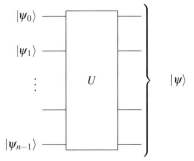

side of Fig. 2.7, is the result of the application of a unitary operator U on the input. The last step is the measurement of the states of each qubit, which returns zeroes and ones that form the final result of the quantum calculation. Note that there is, in principle, an infinite number of possible operators U, which are unitary $2^n \times 2^n$ matrices.

Exercise 2.3 Prove that Hermitean operators have only real eigenvalues.

Exercise 2.4 Prove that unitary operators have only eigenvalues of the form $\exp(i\theta)$, where θ is real.

2.3 Quantum Circuits

The model of quantum circuits is very convenient for representing most of the quantum algorithms currently known.[3] It is analogous to classical digital circuits, with wires conducting the data through logic gates. However, since Nature as described by quantum mechanics is reversible, quantum circuits must be built exclusively from reversible gates. Thus, the model of quantum circuits is more closely related to classical *reversible* circuits. More details on reversible computing may be found in De Vos [15], or in Perumalla [38], and a complete description of quantum circuits can be found in Nielsen and Chuang [36], and in Hirvensalo [19].

Let us start with one-qubit gates. In the classical case there is only one possibility, which is the NOT gate. The NOT gate inverts the bit value: 0 goes to 1 and vice-versa. The straightforward generalization to the quantum case is given in Fig. 2.8, where X is the unitary operator

---

[3]Other influential models are adiabatic quantum computing [16, 44], topological quantum computing [22, 24, 34], cluster-state quantum computing (a.k.a. measurement-based quantum computing or one-way quantum computing) [35, 40], quantum Turing machines [14, 45], quantum random access machines (qRAM) [17, 31], deterministic quantum computing with one qubit (DQC1) [13, 23], IQP computing (a.k.a. temporally unstructured quantum computations) [8, 42], quantum walks [10, 11, 27], and the quantum query model [2, 6], for instance.

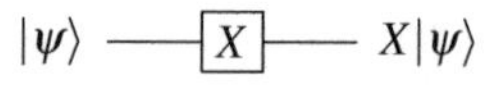

$$|\psi\rangle \quad\boxed{X}\quad X|\psi\rangle$$

Fig. 2.8 Circuit representation of a quantum NOT gate

$$X = \begin{bmatrix} 0 & 1 \\ 1 & 0 \end{bmatrix}. \tag{2.40}$$

Therefore, if the input $|\psi\rangle$ is $|0\rangle$, the output is $|1\rangle$ and vice-versa. However, we can now have a situation with no classical counterpart. The state $|\psi\rangle$ can be a superposition of states $|0\rangle$ and $|1\rangle$. The general case is $|\psi\rangle = \alpha|0\rangle + \beta|1\rangle$ and the corresponding output is $\alpha|1\rangle + \beta|0\rangle$.

The X gate is not the only one-qubit gate. There are infinitely many, since there are an infinite number of 2×2 unitary matrices. The square root of NOT is certainly a curious example without a classical counterpart. It is the quantum gate defined by the unitary matrix

$$\sqrt{X} = \frac{1}{2} \begin{bmatrix} 1+i & 1-i \\ 1-i & 1+i \end{bmatrix}. \tag{2.41}$$

Notice that the above matrix is defined so that $\sqrt{X}\sqrt{X} = X$, meaning that it is halfway from a NOT operation.

In principle, *any* unitary operation is allowed as a quantum gate and could be implemented in hardware, although some operations are more commonly used. The X gate, for instance, is part of a family known as the Pauli gates. Besides the X gate, defined by Eq. (2.40), the Pauli family includes the Y gate, defined by

$$Y = \begin{bmatrix} 0 & -i \\ i & 0 \end{bmatrix}, \tag{2.42}$$

and the Z gate, defined by

$$Z = \begin{bmatrix} 1 & 0 \\ 0 & -1 \end{bmatrix}. \tag{2.43}$$

They may also be represented by the symbols σ_x, σ_y and σ_z, respectively. The Pauli matrices have many interesting properties and are widely used in quantum mechanics. Although most of those properties are out of the scope of this book, some of them are worth mentioning. For instance, we implicitly used Pauli-Z gate earlier in this chapter, when describing a measurement in the computational basis. Moreover, Pauli matrices can be used to define the rotation operators

$$R_X(\theta) = \exp\left(-\frac{i\theta}{2}X\right), \tag{2.44}$$

$$R_Y(\theta) = \exp\left(-\frac{i\theta}{2}Y\right), \tag{2.45}$$

and

$$R_Z(\theta) = \exp\left(-\frac{i\theta}{2}Z\right), \tag{2.46}$$

around the respective axes of the Bloch sphere. See Exercises 2.5 and 2.6 at the end of this section.

Another important family of one-qubit gates is composed by the phase shift gates, defined by

$$P(\theta) = \begin{bmatrix} 1 & 0 \\ 0 & \exp(i\theta) \end{bmatrix}, \tag{2.47}$$

which leaves the basis state $|0\rangle$ unchanged and maps the basis state $|1\rangle$ onto $\exp(i\theta)|1\rangle$. It may seem that the phase shift only changes *global* phases. However, notice that in the general case, the phase shift maps a qubit $\alpha|0\rangle + \beta|1\rangle$ onto $\alpha|0\rangle + \beta\exp(i\theta)|1\rangle$, in which case an important *relative* phase is changed. There are at least two special cases of the phase shift gate that are worth mentioning. One of them is achieved by setting $\theta = \frac{\pi}{4}$, in which case the resulting gate is known as the T gate,[4] with corresponding matrix given by

$$T = \begin{bmatrix} 1 & 0 \\ 0 & \exp(i\frac{\pi}{4}) \end{bmatrix}. \tag{2.48}$$

The other special case is achieved by setting $\theta = \frac{\pi}{2}$, in which case the resulting gate is known as the S gate, with corresponding matrix given by

$$S = \begin{bmatrix} 1 & 0 \\ 0 & i \end{bmatrix}. \tag{2.49}$$

A one-qubit gate can always be decomposed in terms of rotations and a phase shift. Formally, we say that for every 2×2 unitary matrix U, there are real numbers α, β, γ and δ such that

$$U = P(\alpha)R_X(\beta)R_Y(\gamma)R_Z(\delta). \tag{2.50}$$

The proof for the above claim is left as an exercise.

---

[4]T gate is also known in the literature as the $\pi/8$ gate, which may seem a little counterintuitive. The reader should consider that as a special case $P(2\pi/8)$.

The *Hadamard* gate is another important one-qubit gate, given by

$$H = \frac{1}{\sqrt{2}} \begin{bmatrix} 1 & 1 \\ 1 & -1 \end{bmatrix}. \tag{2.51}$$

It is easy to see that

$$H|0\rangle = \frac{|0\rangle + |1\rangle}{\sqrt{2}}, \tag{2.52}$$

$$H|1\rangle = \frac{|0\rangle - |1\rangle}{\sqrt{2}}. \tag{2.53}$$

If the input is $|0\rangle$, the Hadamard gate creates a superposition of states with equal weights. This is a general feature, valid for two or more qubits. Let us analyze the two-qubit case. In order to do that, we must use the postulate of composition and the tensor product.

As our first example of a two-qubit gate, we present $H \otimes H$, which can be applied to input $|0\rangle \otimes |0\rangle$, yielding

$$\begin{aligned} (H \otimes H)|0\rangle \otimes |0\rangle &= H|0\rangle \otimes H|0\rangle \\ &= \left(\frac{|0\rangle + |1\rangle}{\sqrt{2}} \right) \otimes \left(\frac{|0\rangle + |1\rangle}{\sqrt{2}} \right) \\ &= \frac{1}{2}(|0\rangle|0\rangle + |0\rangle|1\rangle + |1\rangle|0\rangle + |1\rangle|1\rangle), \end{aligned} \tag{2.54}$$

that may also be written as

$$(H \otimes H)|0\rangle \otimes |0\rangle = \frac{1}{2}(|0\rangle + |1\rangle + |2\rangle + |3\rangle). \tag{2.55}$$

The result is a superposition of all basis states with equal weights. More generally, the Hadamard operator applied to the n-qubit state $|0\rangle$ is

$$\begin{aligned} H^{\otimes n}|0\rangle &= H^{\otimes n}|0, \dots, 0\rangle \\ &= \left(\frac{|0\rangle + |1\rangle}{\sqrt{2}} \right)^{\otimes n} \\ &= \frac{1}{\sqrt{2^n}} \sum_{i=0}^{2^n - 1} |i\rangle. \end{aligned} \tag{2.56}$$

Thus, the tensor product of n Hadamard operators produces an equally weighted superposition of all computational basis states, when the input is the state $|0\rangle$. This state is useful for applying quantum parallelism, as we will see ahead.

Another important two-qubit quantum gate is the CNOT gate, which we antici-
pated in Sect. 2.1 in the classical setting. Of course, the classical CNOT can only
deal with classical inputs and outputs. The quantum CNOT gate has two input
qubits, the control and the target qubit, respectively, which can be arbitrary quantum
states. The target qubit is flipped only if the control qubit is set to $|1\rangle$, that is,

$$|00\rangle \rightarrow |00\rangle,$$
$$|01\rangle \rightarrow |01\rangle,$$
$$|10\rangle \rightarrow |11\rangle,$$
$$|11\rangle \rightarrow |10\rangle. \tag{2.57}$$

The action of the CNOT gate can also be represented by the mapping

$$|a, b\rangle \rightarrow |a, a \oplus b\rangle, \tag{2.58}$$

where $\oplus$ is addition modulo 2. Now, let us obtain its matrix representation.
Performing the same calculations that yield Eq. (2.25), we have

$$|00\rangle = \begin{bmatrix} 1 \\ 0 \\ 0 \\ 0 \end{bmatrix}, \quad |01\rangle = \begin{bmatrix} 0 \\ 1 \\ 0 \\ 0 \end{bmatrix}, \quad |10\rangle = \begin{bmatrix} 0 \\ 0 \\ 1 \\ 0 \end{bmatrix}, \quad |11\rangle = \begin{bmatrix} 0 \\ 0 \\ 0 \\ 1 \end{bmatrix}. \tag{2.59}$$

Thus, from Eqs. (2.57) and (2.59), it follows that the matrix representation U_{CNOT} of
the CNOT gate is given by

$$U_{\mathrm{CNOT}} = \begin{bmatrix} 1 & 0 & 0 & 0 \\ 0 & 1 & 0 & 0 \\ 0 & 0 & 0 & 1 \\ 0 & 0 & 1 & 0 \end{bmatrix}. \tag{2.60}$$

Figure 2.9 describes the CNOT gate,[5] where $|i\rangle$ is either $|0\rangle$ or $|1\rangle$. The figure
could lead one to think that the output is always nonentangled, but that is not true,

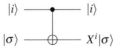

Fig. 2.9 Representation of a CNOT gate. Notice that the control qubit $|i\rangle$, in this case, can be
either $|0\rangle$ or $|1\rangle$. The general case is obtained by linearity

---

[5]Sometimes it may be convenient to have a CNOT with negative control, so that the target qubit is
flipped if and only if the control qubit is $|0\rangle$. A negative control is usually represented in the circuit
model by replacing the full bullet • by an empty bullet ∘ on the control line.

Fig. 2.10 Representation of
a SWAP gate in the circuit
model

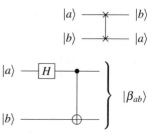

Fig. 2.11 This circuit takes
as input a factorized state
$|a\rangle \otimes |b\rangle$ and outputs one of
the Bell states $|\beta_{ab}\rangle$

since if the first qubit is in a more general state given by $a\,|0\rangle + b\,|1\rangle$, then the
output will be $a\,|0\rangle\,|\sigma\rangle + b\,|1\rangle\,X\,|\sigma\rangle$, which is entangled in general. The input can
be entangled as well. Similarly to the CNOT gate, any other unitary operator U can
be applied conditioned to a control qubit.[6] In that case, operation U is applied to the
target qubit if and only if the control qubit is $|1\rangle$.

The SWAP gate (see Fig. 2.10) is also an important two-qubit quantum gate. It
has two input qubits, and exchanges its values as follows,

$$|00\rangle \rightarrow |00\rangle,$$
$$|01\rangle \rightarrow |10\rangle,$$
$$|10\rangle \rightarrow |01\rangle,$$
$$|11\rangle \rightarrow |11\rangle. \tag{2.61}$$

The matrix representation U_{SWAP} of the SWAP gate is given by

$$U_{\text{SWAP}} = \begin{bmatrix} 1 & 0 & 0 & 0 \\ 0 & 0 & 1 & 0 \\ 0 & 1 & 0 & 0 \\ 0 & 0 & 0 & 1 \end{bmatrix}. \tag{2.62}$$

The action of the SWAP gate can also be achieved by the composition of three
CNOT gates. The proof for this statement is left as an exercise.

We have seen two examples of two-qubit gates. The general case is a 4×4
unitary matrix. Gates that are the direct product of other gates, such as $H \otimes H$,
do not produce entanglement. If the input is nonentangled, the output is not either.
On the other hand, the output of the CNOT gate can be entangled while the input
is nonentangled. For instance, the CNOT gate plays a central role on the circuit
depicted in Fig. 2.11, which takes as input nonentangled states, $|a\rangle \otimes |b\rangle$, and
generates as output a Bell state $|\beta_{ab}\rangle$, which is entangled.

---

[6]Negative controls can also be applied in this case. The notation is the same as the one adopted
in classical and quantum CNOTs, that is, replacing the full bullet • by the empty bullet ∘ on the
negative control lines.

Fig. 2.12 Circuit representation of a measurement in the computational basis. Notice the classical wire after the result

We still need a representation for measurements in the quantum circuit model. We will use the meter symbol for the measurement process, as we already used it for a similar purpose in the classical case. The usual representation is given in Fig. 2.12. Notice that after the measurement we are using a double wire. This is the standard notation for classical wires in the quantum circuit model, i.e., wires that carry only classical bits.

Exercise 2.5 (Spectral Decomposition)   We say that a linear operator U has spectral decomposition if it can be written as $A = \sum_i \lambda_i |a_i\rangle\langle a_i|$, where $|a_i\rangle$ are its eigenvectors, and λ_i are its corresponding eigenvalues. Prove that normal operators always have spectral decomposition.

Exercise 2.6 (Function of Operators)   Given a function f and a linear operator with spectral decomposition $A = \sum_i \lambda_i |a_i\rangle\langle a_i|$, we say that $f(A) = \sum_i f(\lambda_i)|a_i\rangle\langle a_i|$. Find the spectral decomposition of the Pauli matrices X, Y, and Z. Then, find the matrix representation of the rotation matrices $R_X(\theta)$, $R_X(\theta)$, and $R_X(\theta)$, in the computational basis.

Exercise 2.7   Prove that for every 2×2 unitary matrix U, there are real numbers α, β, γ and δ such that

$$U = P(\alpha)R_X(\beta)R_Y(\gamma)R_Z(\delta), \tag{2.63}$$

where matrices P, R_X, R_Y, and R_Z are given by Eqs. (2.44)–(2.47).

Exercise 2.8   Prove that the effect of a SWAP gate can be achieved by the composition of three CNOT gates.

Exercise 2.9   Suppose you want to perform a CNOT with control on qubit a and target on qubit b, but the hardware at your disposal only allows you to perform the operation the other way around, i.e., with control on b and target on a. However, you are also allowed to use other gates from a universal set—including the Hadamard gate. How could you achieve the desired result?

Exercise 2.10   Show that the controlled-Z gate with control on qubit a and target on qubit b performs the same transformation even if we exchange target and control bits, i.e., control on qubit b and target on qubit a.

Exercise 2.11 (No-cloning Theorem)   Prove that an arbitrary quantum bit cannot the copied. *Hint:* Assume that there is a circuit with two input qubits that could be used to copy the state of the first qubit into the second one. Find a contradiction.

Exercise 2.12 (Swap Test) Show that you can use Hadamard gates and a controlled swap gate in order to test whether two quantum states are approximately the same. *Hint:* Notice that two quantum states are the same if their inner product is equal to one.

2.4 Universal Gates

In any model of computation, classical or quantum, it is extremely important to have a small set of logic gates such that they are sufficient for evaluating any computable function. This universal set allows engineers to concentrate on implementing only a few simple operations. If we fail to describe such universal set, or if this set is too complex, then the corresponding model of computation may never become practical.

We already mentioned in Sect. 2.1 that the set consisting of gates AND and NOT is universal for classical computing. In quantum computing, the set consisting of the CNOT gate and *every* one-qubit gate is universal [5]. This means that any other quantum gate, operating on two or more qubits, can be written as compositions and direct products of CNOTs and one-qubit gates. One might argue that this set of gates is not good enough, since it is infinite, and therefore too complex to be implemented. We will come back to that later in this section. Meanwhile, let us present an sketch of the proof that CNOTs and one-qubit gates form a universal set. This proof can be divided into two great steps, as follows.

In the first step, we should prove that an arbitrary unitary operator U can be decomposed into a product of *two-level unitary matrices*, which are unitary matrices that act non-trivially on no more than two vector components. This is left as an exercise, with the following hints. The reader may start from a simpler case, where

$$U = \begin{bmatrix} a & b & c \\ d & e & f \\ g & h & j \end{bmatrix} \tag{2.64}$$

is an arbitrary 3×3 unitary matrix. Then, find a two-level unitary matrix U_1 such that

$$U_1 U = \begin{bmatrix} a' & b' & c' \\ 0 & e' & f' \\ g' & h' & j' \end{bmatrix}. \tag{2.65}$$

Now, find a two-level unitary matrix U_2 such that

$$U_2 U_1 U = \begin{bmatrix} a'' & b'' & c'' \\ 0 & e'' & f'' \\ 0 & h'' & j'' \end{bmatrix}. \tag{2.66}$$

Finally, find a two-level unitary matrix U_3 such that $U_3 U_2 U_1 U = I$. Thus, $U = U_1^\dagger U_2^\dagger U_3^\dagger$. Notice that the adjoint of a two-level unitary matrix is also a two-level unitary matrix. The exercise proposed at the end of the section asks to generalize this procedure taking U as an arbitrary $d \times d$ unitary matrix.

In the second step, we should prove that two-level unitary matrices can be decomposed into CNOTs and one-qubit gates. This is also left as an exercise to the reader, and this time we describe our hints as an example. Let us say that the two-level unitary matrix to be decomposed is of the form

$$U = \begin{bmatrix} 1 & 0 & 0 & 0 & 0 & 0 & 0 & 0 \\ 0 & a & 0 & 0 & 0 & 0 & b & 0 \\ 0 & 0 & 1 & 0 & 0 & 0 & 0 & 0 \\ 0 & 0 & 0 & 1 & 0 & 0 & 0 & 0 \\ 0 & 0 & 0 & 0 & 1 & 0 & 0 & 0 \\ 0 & 0 & 0 & 0 & 0 & 1 & 0 & 0 \\ 0 & c & 0 & 0 & 0 & 0 & d & 0 \\ 0 & 0 & 0 & 0 & 0 & 0 & 0 & 1 \end{bmatrix}. \tag{2.67}$$

Notice that U acts on a system of three qubits, and its non-trivial part corresponds to the operator

$$\tilde{U} = \begin{bmatrix} a & b \\ c & d \end{bmatrix} \tag{2.68}$$

acting on the basis states $|001\rangle$ and $|110\rangle$. Now, we have to write a Gray code—an ordering of binary numbers such that consecutive values differ by exact one bit—connecting 001 to 110. For instance, it could be $001 \longrightarrow 000 \longrightarrow 010 \longrightarrow 110$. With this sequence of binary numbers in mind, it is not difficult to find a sequence of Toffoli gates—possibly with some negative controls—that changes state $|001\rangle$ to $|010\rangle$. Notice that we stopped before the last transition, which in this case would be a bit flip on the most significant bit. Then, we apply a controlled $\tilde{U}$ gate with target on the most significant qubit and controls on all other qubits. Finally, we have to undo the Toffoli gates in order to change $|010\rangle$ back to $|001\rangle$.

Our example is almost done. The problem is that we still have Toffoli gates and a controlled $\tilde{U}$ gate in our decomposition, and they must be reduced to CNOTs and one-qubit gates. Fortunately, this can be easily achieved by the generic process depicted in Fig. 2.13, where U represents an arbitrary unitary operation, and V is a unitary operation such that $U = V^2$. Notice that the decomposition can be applied recursively, in case we have more than two control qubits.

A controlled gate with a single control can also be decomposed directly as in Fig. 2.14, where angle θ and unitary operators A, B, and C must satisfy $ABC = I$ and $U = \exp(i\theta) AXBXC$.

We have just argued that one can decompose an arbitrary unitary operator into a product of two-level unitary matrices, and one can also decompose an arbitrary two-

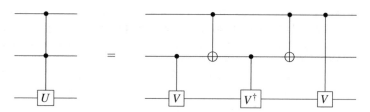

Fig. 2.13 Decomposition of a controlled gate into CNOTs and one-qubit gates

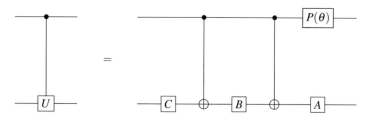

Fig. 2.14 Representation of a controlled-U gate, and its corresponding decomposition in terms of CNOTs and one-qubit gates

level unitary matrix into CNOTs and one-qubit gates. Therefore, the set consisting of the CNOT gate and *every* one-qubit gate is universal for quantum computing. However, as we mentioned before, an infinite universal set is not good enough. The number of possible quantum gates is uncountable, so we need to relax the notion of universality in order to achieve a finite universal set. Instead of requiring exact decompositions in terms of elementary gates, we may allow the quantum operations to be approximated by a sequence of gates from a finite set. In this context, we say that a set of gates is universal if any quantum circuit can be *approximated* with arbitrary accuracy using only gates from that set. Then, it is possible to prove— although beyond the scope of this book—that the set formed by a Hamadard gate and a T gate is universal for 1-qubit circuits. If we add a CNOT gate to this set, we have a universal set of quantum gates for circuits of any size.

For detailed proofs regarding universal quantum gates, refer to Barenco et al. [5], and Kaye, Laflamme and Mosca [21].

Exercise 2.13 Prove that an arbitrary unitary operator U can be decomposed into a product of two-level unitary matrices. Follow the hints given through the text of this section.

Exercise 2.14 Generalize the example presented in this section, and explain how an arbitrary two-level unitary matrix can be decomposed into CNOTs and one-qubit gates.

Exercise 2.15 Can you prove that the circuit below is equivalent to a Toffoli gate?

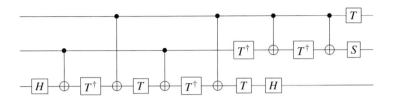

Exercise 2.16 Find a decomposition of a generalized Toffoli gate (that is, a NOT gate with two or more control qubits) in terms of CNOT gates. *Hint:* You will need to add ancillary qubits to the circuit.

2.5 Computational Simulations

Access to a real quantum computer with a large number of qubits is not yet possible nowadays, and we may only perform classical simulations of quantum systems in order to study the behavior of quantum algorithms. These simulations are essential in scenarios where it may be very tedious or impractical to calculate the outcome of the quantum algorithm using only paper and pencil.

It is certainly possible to use one's favorite programming language in order to perform the linear algebra operations studied throughout this chapter. This is particularly convenient for the simulation of basic algorithms with only a few quantum gates. Python have good libraries for quantum computing simulation, such as NumPy for numerical calculations, SymPy for symbolic calculations, and QuTiP for the simulation of open quantum systems. Mathematica, Maple and Matlab are also popular choices. When dealing with more complex quantum algorithms, or when high performance is required, it is usually a better idea to use tools designed specifically for that.

LIQUiD is a software suite for quantum computing simulation designed by Microsoft. It is composed by a programming language based on F#, optimization and scheduling tools, and simulators. Microsoft LIQUiD allows the user to write high-level descriptions of quantum algorithms and translates it to optimized low-level machine instructions of a quantum device in order to simulate them more efficiently. More recently, Microsoft released the Q# language, intended to be a programming language for real quantum devices.[7]

Recently, IBM started to offer the Quantum Experience, granting free access to prototype quantum computers and simulators via the IBM Cloud platform.[8] The first

---

[7]Microsoft LIQUiD can be downloaded from http://stationq.github.io/Liquid. The Microsoft Quantum Development Kit, which includes Q#, can be downloaded from https://www.microsoft.com/en-us/quantum/development-kit.

[8]IBM Quantum Experience can be accessed at https://www.research.ibm.com/ibm-q/.

available device have 5 qubits and can be accessed directly from the Web browser. Algorithms can be expressed in the circuit model through a graphical composer, or can be entered in an assembly language known as QASM. Instead of using the Web browser, there is also the option of using a software package known as QISKit to write quantum programs in Python language with remote access to the IBM quantum devices. Currently, there is a 16-qubit device available to the general public exclusively through QISKit, and a 20-qubit device available to hubs, partners, and members of the IBM Q Network. IBM also announced a 50 qubit prototype that should be available soon.

ProjectQ is an open-source software for quantum computing simulation which allows the user to write quantum algorithms using Python commands.[9] The quantum algorithm written in ProjectQ can be translated to several types of back-ends, such as quantum simulators or even real quantum computers. For instance, it is possible to translate the high-level description of the quantum algorithm into machine instructions for the IBM Quantum Experience platform and run it directly in the cloud.

Rigetti has a development kit for programming hybrid classical-quantum programs using Python language, available at https://www.rigetti.com/forest. The programs written with this kit can be later simulated or run directly in Rigetti's quantum hardware through their Quantum Cloud Services.[10]

Finally, the Quantiki web-site has a full list of classical simulators of quantum algorithms available at https://www.quantiki.org/wiki/list-qc-simulators, and is constantly updated by the community. For a deeper study on quantum computing simulation, the reader may also refer to Viamontes, Markov and Hayes [46], or to Hardy and Steeb [18].

Exercise 2.17 Write quantum programs in IBM Quantum Experience and Microsoft LiQUiD that prepare two quantum states, $|\Psi_A\rangle$ and $|\Psi_B\rangle$, and then run a swap-test between them (see Exercise 2.12). Test your programs with the following input states:

- $|\Psi_A\rangle = \frac{1}{\sqrt{2}}(|0\rangle + |1\rangle)$ and $|\Psi_B\rangle = \frac{1}{\sqrt{2}}(|0\rangle + |1\rangle)$
- $|\Psi_A\rangle = \frac{1}{\sqrt{2}}(|0\rangle + |1\rangle)$ and $|\Psi_B\rangle = \frac{1}{\sqrt{2}}(|0\rangle - |1\rangle)$
- $|\Psi_A\rangle = \frac{1}{\sqrt{2}}(|0\rangle + |1\rangle)$ and $|\Psi_B\rangle = |0\rangle$
- $|\Psi_A\rangle = |1\rangle$ and $|\Psi_B\rangle = |1\rangle$

Exercise 2.18 Implement a generalized Toffoli gate with three control qubits and one target qubit in the IBM Quantum Experience.

---

[9]ProjectQ can be downloaded from http://projectq.ch.

[10]More details at https://www.rigetti.com/qcs.

References

1. Aaronson, S.: Quantum Computing since Democritus. Cambridge University Press, New York (2013)
2. Ambainis, A.: Quantum query algorithms and lower bounds. In: Löwe, B., Piwinger, B., Räsch, T. (eds.) Classical and New Paradigms of Computation and their Complexity Hierarchies, pp. 15–32. Springer, Amsterdam, (2004)
3. Ambainis, A.: Quantum walk algorithm for element distinctness. SIAM J. Comput. **37**(1), 210–239 (2007). https://doi.org/10.1137/S0097539705447311
4. Arora, S., Barak, B.: Computational Complexity: A Modern Approach. Cambridge University Press, New York (2009)
5. Barenco, A., Bennett, C.H., Cleve, R., DiVincenzo, D.P., Margolus, N., Shor, P., Sleator, T., Smolin, J.A., Weinfurter, H.: Elementary gates for quantum computation. Phys. Rev. A **52**(5), 3457–3467 (1995). https://doi.org/10.1103/PhysRevA.52.3457
6. Barnum, H., Saks, M., Szegedy, M.: Quantum query complexity and semi-definite programming. In: Proceedings of 18th IEEE Annual Conference on Computational Complexity, 2003, pp. 179–193 (2003). https://doi.org/10.1109/CCC.2003.1214419
7. Bernstein, E., Vazirani, U.: Quantum complexity theory. SIAM J. Comput. **26**(5), 1411–1473 (1997). https://doi.org/10.1137/S0097539796300921
8. Bremner, M.J., Jozsa, R., Shepherd, D.J.: Classical simulation of commuting quantum computations implies collapse of the polynomial hierarchy. Proc. R. Soc. Lond. A Math. Phys. Eng. Sci. **467**(2126), 459–472 (2011). https://doi.org/10.1098/rspa.2010.0301
9. Chandra, R., Jacobson, N.T., Moussa, J.E., Frankel, S.H., Kais, S.: Quadratic constrained mixed discrete optimization with an adiabatic quantum optimizer. Phys. Rev. A **90**, 012,308 (2014). https://doi.org/10.1103/PhysRevA.90.012308
10. Childs, A.M.: Universal computation by quantum walk. Phys. Rev. Lett. **102**, 180,501 (2009). https://doi.org/10.1103/PhysRevLett.102.180501
11. Childs, A.M., Gosset, D., Webb, Z.: Universal computation by multiparticle quantum walk. Science **339**(6121), 791–794 (2013). https://doi.org/10.1126/science.1229957
12. Cohen-Tannoudji, C., Diu, B., Laloë, F.: Quantum Mechanics. Wiley, Hoboken (1977)
13. Datta, A., Shaji, A., Caves, C.M.: Quantum discord and the power of one qubit. Phys. Rev. Lett. **100**, 050,502 (2008). https://doi.org/10.1103/PhysRevLett.100.050502
14. Deutsch, D.: Quantum theory, the Church-Turing principle and the universal quantum computer. Proc. R. Soc. A Math. Phys. Eng. Sci. **400**(1818), 97–117 (1985). https://doi.org/10.1098/rspa.1985.0070
15. De Vos, A.: Reversible Computing: Fundamentals, Quantum Computing, and Applications. Wiley, Hoboken (2011)
16. Farhi, E., Goldstone, J., Gutmann, S., Sipser, M.: Quantum computation by adiabatic evolution. Technical Report MIT-CTP-2936, Massachusetts Institute of Technology (2000). ArXiv:quant-ph/0001106
17. Gay, S.J.: Quantum programming languages: survey and bibliography. Math. Struct. Comput. Sci. **16**(4), 581–600 (2006). https://doi.org/10.1017/S0960129506005378
18. Hardy, Y., Steeb, W.H.: Classical and Quantum Computing: With C++ and Java Simulations. Birkhäuser, Basel (2001). https://doi.org/10.1007/978-3-0348-8366-5
19. Hirvensalo, M.: Quantum computing. In: Natural Computing Series. Springer, Berlin (2001)
20. Hoffman, K., Kunze, R.A.: Linear Algebra. Prentice-Hall, Upper Saddle River (1971)
21. Kaye, P., Laflamme, R., Mosca, M.: An Introduction to Quantum Computing. Oxford University Press, Oxford (2007)
22. Kitaev, A.: Fault-tolerant quantum computation by anyons. Ann. Phys. **303**(1), 2–30 (2003). https://doi.org/10.1016/S0003-4916(02)00018-0
23. Knill, E., Laflamme, R.: Power of one bit of quantum information. Phys. Rev. Lett. **81**, 5672–5675 (1998). https://doi.org/10.1103/PhysRevLett.81.5672

24. Lahtinen, V., Pachos, J.K.: A short introduction to topological quantum computation. SciPost Phys. **3**, 021 (2017). https://10.21468/SciPostPhys.3.3.021
25. Lang, S.: Linear Algebra. Springer, Heidelberg (1989)
26. Lewis, H.R., Papadimitriou, C.H.: Elements of the Theory of Computation, 2nd edn. Prentice-Hall, Upper Saddle River (1997)
27. Lovett, N.B., Cooper, S., Everitt, M., Trevers, M., Kendon, V.: Universal quantum computation using the discrete-time quantum walk. Phys. Rev. A **81**, 042,330 (2010). https://doi.org/10.1103/PhysRevA.81.042330
28. Magniez, F., Santha, M., Szegedy, M.: Quantum algorithms for the triangle problem. In: Proceedings of the Sixteenth Annual ACM-SIAM Symposium on Discrete Algorithms, SODA '05, pp. 1109–1117. Society for Industrial and Applied Mathematics, Philadelphia (2005)
29. Magniez, F., Nayak, A., Roland, J., Santha, M.: Search via quantum walk. SIAM J. Comput. **40**(1), 142–164 (2011). https://doi.org/10.1137/090745854
30. Messiah, A.: Quantum Mechanics. Dover, Mineola (2014)
31. Miszczak, J.A.: Models of quantum computation and quantum programming languages. Bull. Acad. Pol. Sci. Tech. Sci. **59**(3), 305–324 (2011). https://doi.org/10.2478/v10175-011-0039-5
32. Miszczak, J.A.: High-level Structures for Quantum Computing. Synthesis Lectures on Quantum Computing. Morgan & Claypool, San Rafael (2012)
33. Montanaro, A.: Quantum algorithms: an overview. NPJ Quantum Inf. **2**(15023) (2016). https://doi.org/10.1038/npjqi.2015.23
34. Nayak, C., Simon, S.H., Stern, A., Freedman, M., Das Sarma, S.: Non-abelian anyons and topological quantum computation. Rev. Mod. Phys. **80**, 1083–1159 (2008). https://doi.org/10.1103/RevModPhys.80.1083
35. Nielsen, M.A.: Cluster-state quantum computation. Rep. Math. Phys. **57**(1), 147–161 (2006). https://doi.org/10.1016/S0034-4877(06)80014-5
36. Nielsen, M.A., Chuang, I.L.: Quantum Computation and Quantum Information. Cambridge University, Cambridge (2010). https://doi.org/10.1017/CBO9780511976667
37. Papadimitriou, C.H.: Computational Complexity. Addison-Wesley Pub Co, Boston (1994)
38. Perumalla, K.S.: Introduction to Reversible Computing. Computational Science Series. Chapman & Hall/CRC, Boca Raton (2013)
39. Portugal, R.: Quantum Walks and Search Algorithms. Springer, Cham (2018). https://doi.org/10.1007/978-3-319-97813-0
40. Raussendorf, R., Briegel, H.J.: A one-way quantum computer. Phys. Rev. Lett. **86**, 5188–5191 (2001). https://doi.org/10.1103/PhysRevLett.86.5188
41. Reichardt, B.W.: Reflections for quantum query algorithms. In: Proceedings of the Twenty-second Annual ACM-SIAM Symposium on Discrete Algorithms, SODA '11, pp. 560–569. Society for Industrial and Applied Mathematics, Philadelphia (2011)
42. Shepherd, D., Bremner, M.J.: Temporally unstructured quantum computation. Proc. R. Soc. Lond. A Math. Phys. Eng. Sci. (2009). https://doi.org/10.1098/rspa.2008.0443
43. Shor, P.: Algorithms for quantum computation: discrete logarithms and factoring. In: Proceedings 35th Annual Symposium on Foundations of Computer Science, pp. 124–134. IEEE Computer Society, Washington (1994). https://doi.org/10.1109/SFCS.1994.365700
44. van Dam, W., Mosca, M., Vazirani, U.: How powerful is adiabatic quantum computation? In: Proceedings 2001 IEEE International Conference on Cluster Computing, pp. 279–287 (2001). https://doi.org/10.1109/SFCS.2001.959902
45. Vazirani, U.V.: A survey of quantum complexity theory. In: Lomonaco Jr., S.J. (ed.) Quantum Computation: A Grand Mathematical Challenge for the Twenty-First Century and the Millennium. Proceedings of Symposia in Applied Mathematics, vol. 58, pp. 193–217. American Mathematical Society, Providence (2002)
46. Viamontes, G.F., Markov, I.L., Hayes, J.P.: Quantum Circuit Simulation. Springer, Amsterdam (2009)

Chapter 3
Grover's Algorithm for Unstructured Search

Suppose we have an unstructured database with N elements. Without loss of generality, consider that the elements are labeled as integer numbers from 0 to $N - 1$. The elements are not ordered and we want to find a specific entry in this database. Although the problem of searching for an element on an unsorted list may seem very restrict, keep in mind that many important problems in computer science may be reduced to that. In particular, the classical method of brute-force search can be understood as searching for an element on an unsorted list. If there is no hint about where the searched element is more likely to be found, then any classical algorithm should test each element, one at a time. In the worst case, N trials are necessary to guarantee that the element will be found. As we shall see, by using quantum mechanics, only $\Theta(\sqrt{N})$ trials are needed. In both cases, there should be an efficient way of verifying if the element being tried is the searched element. Fortunately, for any problem in complexity class NP—consequently, even for NP-complete problems—we guarantee that this is possible.

3.1 Introduction

The search algorithm described in this chapter was introduced by Lov Grover in [13]. References [11, 12, 14, 16–18] are also related. In this book, we use the circuit model, so that the presentation is kept close to the original. Alternative formulations based on the Schrödinger equation with continuous-time evolution of the Hamiltonian have also been explored in the literature [8, 9, 22, 25]. Both discrete and continuous time approaches are closely related, and find the marked element in time proportional to the square root of the size of the list [26].

For simplicity, assume that $N = 2^n$, for some integer n. Grover's algorithm requires two registers: n qubits in the first and one qubit in the second. In the first step of the algorithm, we need to create a uniform superposition of all 2^n

© The Author(s), under exclusive license to Springer Nature Switzerland AG 2019
F. de Lima Marquezino et al., *A Primer on Quantum Computing*, SpringerBriefs in
Computer Science, https://doi.org/10.1007/978-3-030-19066-8_3

computational basis states $\{|0\rangle, \ldots, |2^n - 1\rangle\}$ of the first register.[1] This can be achieved by initializing each qubit of the first register in the state $|0\rangle$ and then applying the Hadamard operator on them. That is, we produce the state

$$
\begin{aligned}
|\psi\rangle &= H^{\otimes n} |00 \ldots 0\rangle \\
&= \left(\frac{|0\rangle + |1\rangle}{\sqrt{2}} \right)^{\otimes n} \\
&= \frac{1}{\sqrt{N}} \sum_{i=0}^{N-1} |i\rangle,
\end{aligned}
\tag{3.1}
$$

a superposition of all basis states with equal amplitudes given by $1/\sqrt{N}$. The single qubit from the second register may start in state $|1\rangle$ and, after applying a Hadamard operator to it, we achieve the state

$$
|-\rangle = \frac{|0\rangle - |1\rangle}{\sqrt{2}}.
\tag{3.2}
$$

Now, we define $f : \{0, \ldots, N - 1\} \to \{0, 1\}$ as a function which recognizes the solution.[2] If the searched element is labeled as i_0, then

$$
f(i) = \begin{cases} 1, & \text{if } i = i_0 \\ 0, & \text{otherwise.} \end{cases}
\tag{3.3}
$$

Notice that, by definition, any problem in complexity class[3] NP allows such a function to be efficiently evaluated by a classical algorithm. In the quantum setting, we should build a linear unitary operator U_f, also dependent on f, such that

$$
U_f |i\rangle |j\rangle = |i\rangle |j \oplus f(i)\rangle.
\tag{3.4}
$$

We call U_f an *oracle*. In the above equation, $|i\rangle$ stands for a state of the first register, so i is in $\{0, \ldots, 2^n - 1\}$, $|j\rangle$ is a state of the second register, so j is in $\{0, 1\}$, and the sum is modulo two. It is easy to check that

---

[1] Biron et al. generalized Grover's algorithm to work with arbitrary initial amplitude distributions [4]. In this book, we assume the uniform superposition in the input for simplicity.

[2] For simplicity, we assume that there is only one solution. However, Boyer, Brassard, Høyer, and Tapp have already generalized Grover's algorithm for searching with repeated elements. They have also presented a first version of a *counting algorithm* [5]. Mosca also described a version of the counting algorithm, using a different approach called *phase estimation* [21].

[3] Arora and Barak's textbook on computational complexity discusses Grover's algorithm and its relation to complexity classes [2].

$$U_f \, |i\rangle \, |-\rangle = \frac{U_f \, |i\rangle \, |0\rangle - U_f \, |i\rangle \, |1\rangle}{\sqrt{2}}$$

$$= \frac{|i\rangle \, |f(i)\rangle - |i\rangle \, |1 \oplus f(i)\rangle}{\sqrt{2}}$$

$$= (-1)^{f(i)} \, |i\rangle \, |-\rangle . \tag{3.5}$$

In the last equality, we have used the fact that

$$1 \oplus f(i) = \begin{cases} 0, & \text{for } i = i_0 \\ 1, & \text{for } i \neq i_0. \end{cases} \tag{3.6}$$

Now, look at what happens when we apply U_f to the superposition state coming from the first step, $|\psi\rangle \, |-\rangle$. The state of the second register does not change. Let us call $|\psi_1\rangle$ the resulting state of the first register. Then, we have

$$|\psi_1\rangle \, |-\rangle = U_f \, |\psi\rangle \, |-\rangle$$

$$= \frac{1}{\sqrt{N}} \sum_{i=0}^{N-1} U_f \, |i\rangle \, |-\rangle$$

$$= \frac{1}{\sqrt{N}} \sum_{i=0}^{N-1} (-1)^{f(i)} \, |i\rangle \, |-\rangle . \tag{3.7}$$

Notice that $|\psi_1\rangle$ is a superposition of all basis elements, but the amplitude of the searched element is negative while all others are positive. The searched element has been marked with a minus sign. This result is obtained using a feature called *quantum parallelism*. At the quantum level, it is possible to check all database elements simultaneously. The position of the searched element is known: it is the value of i of the term with negative amplitude in Eq. (3.7). However, this quantum information is not fully available at the classical level. A classical information of a quantum state is obtained by practical measurements, and, at this point, it does not help if we measure the state of the first register, because it is much more likely that we obtain a non-desired element instead of the searched one. Before we can perform a measurement, the next step should be to increase the amplitude of the searched element while decreasing the amplitude of the others. This is quite general: quantum algorithms work by increasing the amplitude of the states which carry the desired result. After that, a measurement will hit the solution with high probability.

Now we shall work out the details by introducing the circuit for Grover's algorithm (Fig. 3.1) and analyzing it step by step. Notice that unitary operator G is applied $O(\sqrt{N})$ times. The exact number will be obtained later on. The circuit for one Grover iteration G is detailed in Fig. 3.2.

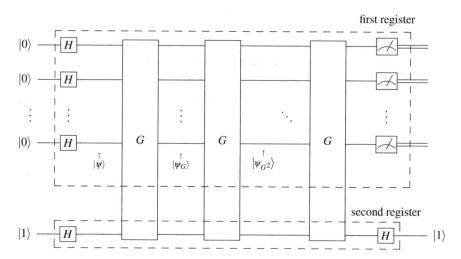

Fig. 3.1 Outline of Grover's algorithm. The fist register has n qubits, and the second register has only one qubit. The Grover operator G is repeated $O(\sqrt{2^n})$ times

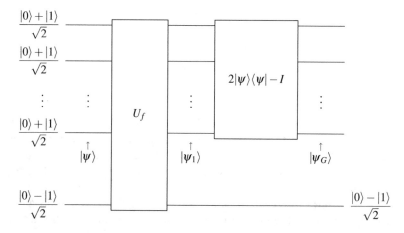

Fig. 3.2 One Grover iteration (G). The states of the first register correspond to the first iteration

The states $|\psi\rangle$ and $|\psi_1\rangle$ are given by Eqs. (3.1) and (3.7), respectively. The operator $2|\psi\rangle\langle\psi| - I$ is called inversion about the mean for reasons that will be clear in the next section. We will also show how each Grover operator application raises the amplitude of the searched element. State $|\psi_1\rangle$ can be rewritten as

$$|\psi_1\rangle = |\psi\rangle - \frac{2}{\sqrt{2^n}}|i_0\rangle, \tag{3.8}$$

where $|i_0\rangle$ is the searched element and is a state of the computational basis. Notice that

$$\langle \psi | i_0 \rangle = \frac{1}{\sqrt{2^n}}. \tag{3.9}$$

Let us calculate $|\psi_G\rangle$ of Fig. 3.2. By using Eqs. (3.8) and (3.9), we achieve

$$|\psi_G\rangle = (2 |\psi\rangle \langle \psi| - I) |\psi_1\rangle$$

$$= \frac{2^{n-2} - 1}{2^{n-2}} |\psi\rangle + \frac{2}{\sqrt{2^n}} |i_0\rangle. \tag{3.10}$$

This is the state of the first register after one application of G. The second register is in state $|-\rangle$.

Exercise 3.1 Show the general form of the entries of the matrix representation for U_f in the computational basis, for an arbitrary input size N and arbitrary searched element i_0.

Exercise 3.2 Show the general form of the entries of the matrix representation for the inversion about the mean operator in the computational basis, for an arbitrary input size N.

Exercise 3.3 Show the general form of the entries of the matrix representation for Grover operator G in the computational basis, for an arbitrary input size N and arbitrary searched element i_0.

Exercise 3.4 Find the eigenvalues and eigenvectors of the Grover operator G.

Exercise 3.5 Express state $|\psi\rangle = \frac{1}{\sqrt{N}} \sum_{i=0}^{N-1} |N\rangle$ in the eigenbasis of the Grover operator U. Use this result to find the minimum value of t_f such that $U^{t_f} |\psi\rangle$ is orthogonal to $|\psi\rangle$.

3.2 Geometric Visualization

All the operators and amplitudes in Grover's algorithm are real. This means that during the execution of this algorithm, all states of the quantum computer will be in a real vector subspace of the Hilbert space. This allows a nice geometrical representation taking $|i_0\rangle$ and $|\psi\rangle$ as non-orthogonal base vectors.

In Fig. 3.3, we have the vectors $|i_0\rangle$ and $|\psi\rangle$. They form an angle smaller than $90°$, according to Eq. (3.9), since $0 < \langle \psi | i_0 \rangle < 1$. If n is large, then the angle is nearly $90°$. In this representation, $|\psi\rangle$ is the initial state of the first register, and the steps of the computation correspond to alternate applications of the oracle operator and the operator of inversion about the mean. Then, state of the quantum computer will be rotated in the real plane spanned by $|\psi\rangle$ and $|i_0\rangle$, keeping the unit norm. This means that the state vector remains in the unit circle during every step of the algorithm.

Fig. 3.3 The state of the first register lives in the real vector space spanned by $|i_0\rangle$ and $|\psi\rangle$. We take these states as a basis to describe what happens in Grover's algorithm

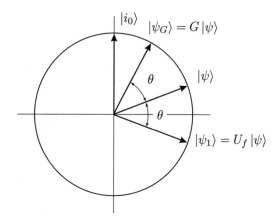

From Eqs. (3.8) and (3.9), we find that $|\psi\rangle$ is rotated by an angle of θ degrees clockwise, where

$$\cos\theta = 1 - \frac{1}{2^{n-1}}. \tag{3.11}$$

This is represented as the angle between states $|\psi\rangle$ and $|\psi_1\rangle$ in Fig. 3.3. From Eq. (3.10), we also find that the angle between $|\psi_G\rangle$ and $|\psi\rangle$ is given by

$$\cos\theta' = \langle\psi|\psi_G\rangle$$

$$= 1 - \frac{1}{2^{n-1}}. \tag{3.12}$$

Hence, we have $\theta' = \theta$ and state $|\psi_1\rangle$ is rotated by an angle of 2θ degrees counterclockwise (in the direction of $|i_0\rangle$). This explains the placement of $|\psi_G\rangle$ in Fig. 3.3. This is a remarkable result, since the resulting action of $G = (2|\psi\rangle\langle\psi| - I)U_f$ rotates $|\psi\rangle$ towards $|i_0\rangle$ by θ degrees. This means that the amplitude of state $|i_0\rangle$ in $|\psi_G\rangle$ increased and the amplitudes of states $|i \neq i_0\rangle$ decreased with respect to their original values in $|\psi\rangle$. A measurement, at this point, will return $|i_0\rangle$ more likely than before. But that is not enough in general—notice, from Eq. (3.11), that θ is a small angle when $n \gg 1$. That is why we need to apply G repeatedly, ending up θ degrees closer to $|i_0\rangle$ each time, until the state of the first register be very close to $|i_0\rangle$, so we can measure.

Now we show that further applications of G also rotate the state of the first register by θ degrees towards $|i_0\rangle$. The proof is quite general. Suppose that $|\sigma\rangle$ is a unit vector making an angle α_1 with $|\psi\rangle$, as in Fig. 3.4.

Let $|\sigma_1\rangle$ be the state of the first register after the application of U_f on $|\sigma\rangle|-\rangle$. Notice that U_f changes the sign of the component of $|\sigma\rangle$ in the direction of $|i_0\rangle$. Therefore, $|\sigma_1\rangle$ is the reflection of $|\sigma\rangle$ around the horizontal axis. Let α_2 be the angle between $|\psi\rangle$ and $|\sigma_1\rangle$.

Fig. 3.4 A generic vector $|\sigma\rangle$ is reflected around the horizontal axis by the application of U_f, yielding $|\sigma_1\rangle$. Then, the reflection of $|\sigma_1\rangle$ about the mean $|\psi\rangle$ gives $G|\sigma\rangle$, which is θ degrees closer to $|i_0\rangle$ (vertical axis)

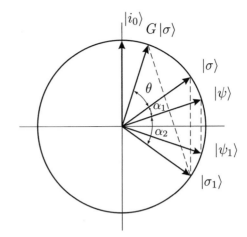

We should prove that $G|\sigma\rangle$ lies in the subspace spanned by $|i_0\rangle$ and $|\psi\rangle$. We have that

$$
\begin{aligned}
G|\sigma\rangle &= (2|\psi\rangle\langle\psi| - I) U_f |\sigma\rangle \\
&= 2\langle\psi|U_f|\sigma\rangle|\psi\rangle - |\sigma_1\rangle \\
&= 2\cos\alpha_2|\psi\rangle - |\sigma_1\rangle.
\end{aligned}
\tag{3.13}
$$

For simplicity, we have omitted the state $|-\rangle$ of the second register in the above calculation. Since $|\sigma_1\rangle$ lies in the subspace spanned by $|i_0\rangle$ and $|\psi\rangle$, we have that $G|\psi\rangle$ also does, as we wanted to prove.

Now we should prove that the angle between $|\sigma\rangle$ and $G|\sigma\rangle$ is equal to θ, which is the same angle between $|\psi\rangle$ and $|\psi_1\rangle$. Notice that

$$
\begin{aligned}
\langle\sigma|G|\sigma\rangle &= 2\langle\sigma|\psi\rangle\cos\alpha_2 - \langle\sigma|\sigma_1\rangle \\
&= \cos\alpha_1\cos\alpha_2 - \cos(\alpha_1 + \alpha_2) \\
&= \cos(\alpha_2 - \alpha_1).
\end{aligned}
\tag{3.14}
$$

From Fig. 3.4, we see that $\alpha_2 - \alpha_1$ is θ. From Eq. (3.13), we see that $G|\sigma\rangle$ is a rotation of $|\sigma\rangle$ towards $|i_0\rangle$ by an angle of θ degrees.

The geometrical interpretation of the operator $2|\psi\rangle\langle\psi| - I$ is that it reflects any real vector around the axis defined by the vector $|\psi\rangle$. For example, in Fig. 3.4, we see that $G|\sigma\rangle = (2|\psi\rangle\langle\psi| - I)|\sigma_1\rangle$ is the reflection of $|\sigma_1\rangle$ around $|\psi\rangle$. The operator $2|\psi\rangle\langle\psi| - I$ is called inversion about the mean for the following reason. Let $|\sigma\rangle = \sum_{i=0}^{2^n-1} \sigma_i|i\rangle$ be an arbitrary vector and take $\langle\sigma\rangle = \sum_{i=0}^{2^n-1} \sigma_i$ as the mean of the amplitudes of $|\sigma\rangle$. If we define

$$|\sigma'\rangle = \sum_{i=0}^{2^n-1} (\sigma_i - \langle\sigma\rangle)\, |i\rangle, \tag{3.15}$$

then we have that

$$(2\,|\psi\rangle\,\langle\psi| - I)\,|\sigma'\rangle = -\,|\sigma'\rangle. \tag{3.16}$$

The above equation shows that a vector with amplitudes $\sigma_i - \langle\sigma\rangle$ is transformed into a vector with amplitudes $-(\sigma_i - \langle\sigma\rangle)$. Notice that $|\sigma'\rangle$ is not normalized, but this is irrelevant in the above argument because the amplitudes of $|\sigma\rangle$ and $|\sigma'\rangle$ only differ by a minus sign.

The oracle operator U_f also has a geometric interpretation, which can be seen from the expression

$$U_f = I - 2\,|i_0\rangle\,\langle i_0|, \tag{3.17}$$

which yields

$$U_f\,|i\rangle = \begin{cases} |i\rangle, & \text{if } i \neq i_0 \\ -\,|i_0\rangle, & \text{if } i = i_0. \end{cases} \tag{3.18}$$

Therefore, the above representation for U_f is equivalent to Eq. (3.5) if we do not consider the state of the second register. The geometric interpretation is that U_f reflects a generic vector about the plane orthogonal to $|i_0\rangle$. This is what Eq. (3.18) shows for vectors of the computational basis. The interpretation is valid for an arbitrary vector because U_f is linear. Since we do not know i_0 before running the algorithm, we had not used Eq. (3.17) to define U_f in the beginning of the description of the algorithm. However, it is perfectly reasonable to assume that function f, as given by Eq. (3.3), can be used to build an efficient implementation of U_f, as given by Eq. (3.4).

3.3   A Small Example for Grover's Algorithm

In this section, we describe Grover's algorithm for a search space of eight elements. If $N = 8$, then $n = 3$. Therefore, we should have three qubits in the first register and one qubit in the second register. For $N = 8$, the operator G will be applied two times—we will see later on how to calculate the optimal number of iterations given the size of the database. The circuit, in this case, is given in Fig. 3.5. The oracle is queried two times. Classically, an average of more than four queries are needed in order to find the searched element with probability at least $1/2$.

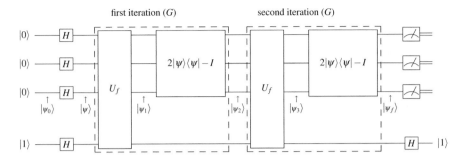

Fig. 3.5 Circuit for Grover's algorithm over an unstructured database of $N = 8$ elements

Let us describe the quantum computer state at each step shown in the circuit—from the initial state $|\psi_0\rangle$ up to the final state $|\psi_f\rangle$. Initially, we have the state

$$|\psi_0\rangle = |000\rangle .\tag{3.19}$$

After applying the Hadamard gates, we get

$$|\psi\rangle = H^{\otimes 3} |000\rangle$$

$$= \frac{1}{2\sqrt{2}} \sum_{i=0}^{7} |i\rangle .\tag{3.20}$$

At this point, we should define the oracle operator that will be used in our example. Of course, in real applications of Grover's algorithm, the oracle operator can be very sophisticated. For instance, it could be marking the state corresponding to the solution of a NP problem. In this case, it would be implemented in a similar way as classical algorithms implement certificate verifiers. These complications, however, can be abstracted from our small example. For simplicity, suppose that we are searching for the element with index 5—which corresponds to the element 101 in binary representation. Therefore, the oracle operator should mark element $|101\rangle$. That is,

$$U_f |101\rangle |-\rangle = -|101\rangle |-\rangle$$
$$U_f |i\rangle |-\rangle = |i\rangle |-\rangle \ , \text{if } i \neq 5.\tag{3.21}$$

After defining the searched state, we should define a vector orthogonal to it, namely

$$|u\rangle = \frac{|000\rangle + |001\rangle + |010\rangle + |011\rangle + |100\rangle + |110\rangle + |111\rangle}{\sqrt{7}}.\tag{3.22}$$

Fig. 3.6 Intermediate states in Grover's algorithm for $N = 8$. Notice how close is $|\psi_f\rangle$ to $|101\rangle$, indicating a high probability that a measurement will give the searched element. The value of θ is around 41.4°

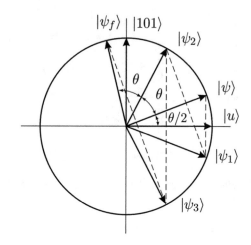

Then, we have that

$$|\psi\rangle = \frac{\sqrt{7}}{2\sqrt{2}} |u\rangle + \frac{1}{2\sqrt{2}} |101\rangle . \qquad (3.23)$$

With this result we can already check the direction of $|\psi\rangle$ in Fig. 3.6. Notice that the value of θ is given by

$$\theta = 2 \arccos \left(\frac{\sqrt{7}}{2\sqrt{2}} \right) , \qquad (3.24)$$

which is approximately 41.4°.

The next step consists in applying the oracle operator, $U_f |\psi_1\rangle |-\rangle$, which produces the result

$$|\psi_1\rangle |-\rangle = \frac{1}{2\sqrt{2}} (|000\rangle + |001\rangle + |010\rangle + |011\rangle +$$

$$+ |100\rangle - |101\rangle + |110\rangle + |111\rangle) |-\rangle , \qquad (3.25)$$

such that $|101\rangle$ is the only term with a minus sign. It is now convenient to rewrite $|\psi_1\rangle$ in order to emphasize this term, as in

$$|\psi_1\rangle = |\psi\rangle - \frac{1}{\sqrt{2}} |101\rangle \qquad (3.26)$$

or

$$|\psi_1\rangle = \frac{\sqrt{7}}{2\sqrt{2}}|u\rangle - \frac{1}{2\sqrt{2}}|101\rangle. \tag{3.27}$$

The form of Eq. (3.26) is useful in the next step of calculation, since we have to apply $2|\psi\rangle\langle\psi| - I$. The form of Eq. (3.27) is useful to draw $|\psi_1\rangle$ in Fig. 3.6, since it allows us to see the state $|\psi_1\rangle$ as the reflection of $|\psi\rangle$ with respect to $|u\rangle$.

In the next step, we apply once again the reflection around the mean, obtaining

$$|\psi_2\rangle = (2|\psi\rangle\langle\psi| - I)|\psi_1\rangle. \tag{3.28}$$

Using Eq. (3.26), we get

$$|\psi_2\rangle = \frac{1}{2}|\psi\rangle + \frac{1}{\sqrt{2}}|101\rangle \tag{3.29}$$

and, using Eq. (3.23), we get

$$|\psi_2\rangle = \frac{\sqrt{7}}{4\sqrt{2}}|u\rangle + \frac{5}{4\sqrt{2}}|101\rangle. \tag{3.30}$$

In order to confirm that the angle between $|\psi\rangle$ and $|\psi_2\rangle$ is θ, notice that

$$\cos\theta = \langle\psi|\psi_2\rangle$$
$$= \frac{3}{4}, \tag{3.31}$$

which agrees with Eq. (3.24). This completes the first application of G, the iteration operator.

The second and last application of G is similar. The next state in our analysis is $|\psi_3\rangle$, which is obtained by the application of the oracle operator, resulting in

$$|\psi_3\rangle = \frac{\sqrt{7}}{2\sqrt{2}}|u\rangle - \frac{5}{4\sqrt{2}}|101\rangle. \tag{3.32}$$

Using Eq. (3.23), we have

$$|\psi_3\rangle = \frac{1}{2}|\psi\rangle - \frac{3}{2\sqrt{2}}|101\rangle. \tag{3.33}$$

Notice that $|\psi_3\rangle$ is the reflection of $|\psi_2\rangle$ with respect to $|u\rangle$.

Finally, the last step consists on applying the inversion around the mean, resulting in the state

$$|\psi_f\rangle = (2|\psi\rangle\langle\psi| - I)|\psi_3\rangle. \tag{3.34}$$

Using Eqs. (3.23) and (3.33), we have

$$|\psi_f\rangle = -\frac{\sqrt{7}}{8\sqrt{2}}|u\rangle + \frac{11}{8\sqrt{2}}|101\rangle.$$ (3.35)

It is easy to confirm that $|\psi_f\rangle$ and $|\psi_2\rangle$ form an angle θ. Notice that the amplitude of the state $|101\rangle$ is much bigger than the amplitude of any other state $|i\rangle$ ($i \neq 5$) in Eq. (3.35). This is the way most quantum algorithms work. They increase the amplitude of the states that carry the desired information. A measurement of the state $|\psi_f\rangle$ in the computational basis will project it into the state $|101\rangle$ with probability

$$p = \left|\frac{11}{8\sqrt{2}}\right|^2 \approx 0.945.$$ (3.36)

Therefore, after two applications of the iteration operator G the chance of getting the result $|101\rangle$, which reads as number 5, is around 94, 5%. Notice that the success probability of Grover's algorithm is not exactly one—even though it is high enough. Nevertheless, Long introduced a modification in Grover's algorithm so that the marked element can be found with probability exactly 1 [20].

Exercise 3.6 Find the circuit for an oracle operator that takes as input $n = 4$ qubits in the first register and yields output one if and only if the input is 1010 or 0101.

Exercise 3.7 Show all the intermediate states of Grover's algorithm for an input list of $N = 16$ elements. Use the oracle operator designed in Exercise 3.6. Notice that there are two searched elements in this case. How does it influence the running time of the algorithm?

3.4 Generalization

The easiest way to calculate the output of Grover's Algorithm is to consider only the action of G instead of breaking the calculation into the action of the oracle and the inversion about the mean. To this end, we choose $|i_0\rangle$ and $|u\rangle$ as the basis for the subspace where $|\psi\rangle$ is rotated after successive applications of G. The searched state is $|i_0\rangle$ and $|u\rangle$ is defined analogously to Eq. (3.22), which in general can be written as

$$|u\rangle = \frac{1}{\sqrt{N-1}} \sum_{\substack{i=0 \\ i \neq i_0}}^{N-1} |i\rangle$$

$$= \sqrt{\frac{N}{N-1}} |\psi\rangle - \frac{1}{\sqrt{N-1}} |i_0\rangle. \tag{3.37}$$

From the first equation above, we find that $|i_0\rangle$ and $|u\rangle$ are orthogonal. From the second equation, we have

$$|\psi\rangle = \sqrt{1 - \frac{1}{N}} |u\rangle + \frac{1}{\sqrt{N}} |i_0\rangle. \tag{3.38}$$

The state of the quantum computer at each step is given by

$$G^k |\psi\rangle = \cos\left(\frac{2k+1}{2}\theta\right) |u\rangle + \sin\left(\frac{2k+1}{2}\theta\right) |i_0\rangle, \tag{3.39}$$

where we have dropped the state of the second register since it is $|-\rangle$ all the time. Eq. (3.39) is obtained from Fig. 3.7 after analyzing the components of $G^k |\psi\rangle$. The value of θ is obtained by replacing k for 0 in Eq. (3.39) and comparing it with Eq. (3.38),

$$\theta = 2 \arccos \sqrt{1 - \frac{1}{N}}. \tag{3.40}$$

Equation (3.39) expresses the fact we proved in Sect. 3.2 that each application of G rotates the state of the first register by θ degrees towards $|i_0\rangle$. Figure 3.7 shows successive applications of G.

The number of times k_0 that G must be applied obeys the equation

$$k_0\theta + \frac{\theta}{2} = \frac{\pi}{2}. \tag{3.41}$$

Fig. 3.7 Effect of operator G applied on state $|\psi\rangle$. Notice that the state of the first register is rotated towards $|i_0\rangle$

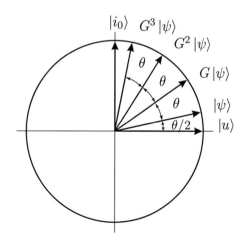

Since k_0 must be integer, we write

$$k_0 = \text{round}\left(\frac{\pi - \theta}{2\theta}\right),\tag{3.42}$$

where θ is given by Eq. (3.40). If $N \gg 1$, by Taylor expanding Eq. (3.40), we get $\theta \approx 2/\sqrt{N}$, and from Eq. (3.42),

$$k_0 = \text{round}\left(\frac{\pi}{4}\sqrt{N}\right).\tag{3.43}$$

After repeating k_0 times the application of G, the probability p of finding the desired element after a measurement is given by

$$p = \sin^2\left(\frac{2k_0 + 1}{2}\theta\right).\tag{3.44}$$

This result is always a rational number. Using Eq. (3.40), we can rewrite probability p as

$$p = \sin^2\left((2k_0 + 1)\arcsin\frac{1}{\sqrt{N}}\right),\tag{3.45}$$

which can be put in terms of the Chebyshev polynomials T as

$$p = T\left(2k_0 + 1, \frac{1}{\sqrt{N}}\right)^2.\tag{3.46}$$

Expanding the Chebyshev polynomials, we obtain

$$p = N\left(\sum_{j=0}^{k_0+1}(-1)^j\binom{2k_0 + 1}{2j - 1}\left(1 - \frac{1}{N}\right)^{k_0+1}\frac{1}{(N - 1)^j}\right)^2,\tag{3.47}$$

where $\binom{a}{b}$ is the binomial coefficient. After replacing N by a power of 2, we can see from the last expression that p is a rational number.

Figure 3.8 shows the success probability p as a function of n, for inputs varying from $n = 2$ to 30. Recall that $N = 2^n$, so for $n = 30$ the search space has around one billion elements.

Notice that when $n = 2$, the probability of getting the result is exactly one. The reason is that Eq. (3.40) yields $\theta = \pi/3$ and $|\psi\rangle$ makes an angle $\pi/6$ with $|u\rangle$. Applying G one time rotates $|\psi\rangle$ to $|i_0\rangle$ exactly. For $n = 3$, Eq. (3.44) yields $p \approx 0.945$, which is the result obtained in Eq. (3.36) from the previous section.

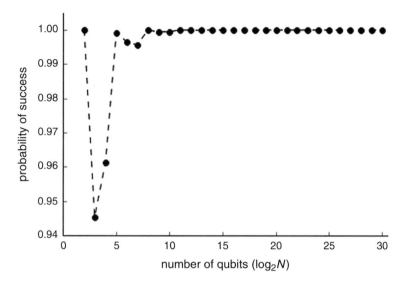

Fig. 3.8 Probability of succeeding as a function of n

3.5   Grover Operator in Terms of Universal Gates

Before running a quantum algorithm on a real quantum computer, it will be necessary to decompose the unitary operator in terms of a universal set. When dealing with complex algorithms, we may even need the aid of a software to automatically decompose the operator for us. Something similar happens in classical computation, where we usually need a compiler to convert the source code written in a high level programming language into very basic binary instructions. In this section, we give an overview on how to decompose the Grover operator G in terms of universal gates, which are CNOT and a small set of one-qubit gates. This decomposition shows how to implement G in practice.

Let us begin by decomposing the inversion about the mean, $2 |\psi\rangle \langle\psi| - I$. Recall that

$$|\psi\rangle = H^{\otimes n} |0\rangle . \tag{3.48}$$

Then,

$$2 |\psi\rangle \langle\psi| - I = H^{\otimes n} (2 |0\rangle \langle0| - I) H^{\otimes n} . \tag{3.49}$$

This equation shows that the central problem is actually to decompose operator $2 |0\rangle \langle0| - I$, which inverts a generic vector about the vector $|0\rangle$. The circuit for it is given in Fig. 3.9.

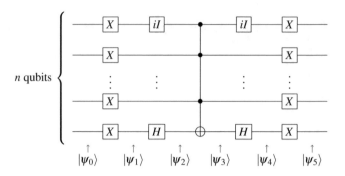

Fig. 3.9 Circuit for $2\,|0\rangle\,\langle 0| - I$. When implementing this circuit, gates iI can be eliminated because they would produce a wrong global phase, which plays no role in the output

One can be convinced that the circuit gives the correct output by following what happens to each state of the computational basis. The input $|0\rangle$ is the only one that does not change signal. The intermediate states as shown in Fig. 3.9 are

$$
\begin{aligned}
|\psi_0\rangle &= & |0\rangle\,|0\rangle\,\ldots\,|0\rangle\,|0\rangle \\
|\psi_1\rangle &= & |1\rangle\,|1\rangle\,\ldots\,|1\rangle\,|1\rangle \\
|\psi_2\rangle &= & i\,|1\rangle\,|1\rangle\,\ldots\,|1\rangle\,|-\rangle \\
|\psi_3\rangle &= & i\,|1\rangle\,|1\rangle\,\ldots\,|1\rangle\,(-\,|-\rangle) \\
|\psi_4\rangle &= -i\,(i\,|1\rangle)\,|1\rangle\,\ldots\,|1\rangle\,|1\rangle \\
|\psi_5\rangle &= & |0\rangle\,|0\rangle\,\ldots\,|0\rangle\,|0\rangle\,.
\end{aligned}
\tag{3.50}
$$

The same calculations for the input $|j\rangle$, with $0 < j < N$, results in $-|j\rangle$ as output. The only operator in Fig. 3.9 that does not act on single qubits is the generalized Toffoli gate. In this case, we can apply the techniques already discussed in Sect. 2.4.

By now, one should be asking about the decomposition of U_f in terms of elementary gates. However, the oracle operator U_f has a different nature from other operators in Grover's algorithm, since its implementation depends on several factors. In the first place, recall that the problem of searching on an unstructured list is actually very broad and can be applied to several problems in NP. Thus, the implementation of the oracle operator may depend on the specific problem being addressed by Grover's algorithm. Moreover, the elements of the input list may be only labels pointing to data which are stored elsewhere—such as primary keys on a relational database. In this case, the implementation of the oracle operator would depend on how data is loaded from a quantum memory of a quantum computer.

On the other hand, we have pointed out that U_f can be represented by $I - 2\,|i_0\rangle\,\langle i_0|$ (Eq. (3.17)), if one knows *a priori* the answer i_0 of the problem. This representation is useful for simulating Grover's algorithm in a classical computer to test its efficiency. The operator $I - 2\,|i_0\rangle\,\langle i_0|$ is decomposed as a generalized Toffoli gate with n control qubits, one target qubit in the state $|-\rangle$, and two symmetrical X

Fig. 3.10 Decomposition of
$I - 2|101\rangle\langle101|$, which
simulates an oracle U_f that
searches for number 5

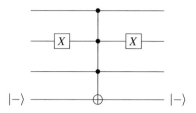

gates in the ith qubit, if the ith binary digit of i_0 is 0. For example, the oracle operator
used in our small example from Sect. 3.3—recall Eq. (3.21)—could be implemented
by the circuit in Fig. 3.10.

In Chap. 2, we have pointed out that the efficiency of an algorithm is measured by
how the number of elementary gates increases as a function of the number of qubits.
If we count the elementary gates from this section and apply Eq. (3.43), we find that
a total of $\pi(17n - 15)\sqrt{2^n} + n + 2$ universal gates were used to implement Grover's
algorithm. This yields a complexity of $O(n\sqrt{2^n})$, or equivalently $\tilde{O}(\sqrt{2^n})$.[4]

It is possible to prove that $\Omega(\sqrt{N})$ is the lower bound for quantum search algorithms on unstructured lists, meaning that it is impossible to find an asymptotically
faster algorithm. Quantum computers will not be more efficient than that if we need
to try every possible solution. Therefore, in order to find quantum algorithms with
better speed-ups than quadratic, we must explore the structure of the input data [1].
In the following chapters, we will show some notable examples of how that can be
achieved. The detailed proofs of complexity for Grover's algorithm are beyond the
scope of this book. Bennett, Bernstein, Brassard, and Vazirani proved the optimality
of Grover's algorithm [3]. For alternative proofs see the textbooks by Portugal [24]
or Nielsen and Chuang [23]. The proof presented by Zalka is more detailed [28].

Grover's algorithm have been implemented several times. Jones, Mosca and
Hansen implemented a small instance of Grover's algorithm—searching for one
out of four possible states—using the technique of nuclear magnetic resonance
(NMR) [15]. Vandersypen and Steffen also used NMR to implement Grover's
algorithm on a three-qubit computer [27]. Brickman et al. implemented a small
instance of Grover's algorithm using the technique of trapped ions [6]. Leuenberger
and Loss proposed an implementation of Grover's algorithm using molecular
magnets [19]. Zhang et al. implemented quantum search using 3D printed metamaterials [29]. Recently, Godfrin et al. also implemented Grover's algorithm using
nuclear spin [10].

---

[4]The notation $\tilde{O}(N)$ means $O(N \log N)$.

3.6   Computational Simulations

Grover's algorithm can be easily simulated by a Python script. It is convenient to import the *NumPy* library for numerical processing. We should also import *matplotlib* if we wish to plot the results. We can do that by the following commands:

```
1  import numpy as np
2  import matplotlib . pyplot  as  plt
```

$$(3.51)$$

Now, we define the size of the input list and the searched element, simply by initializing some variables:

```
1  n = 8
2  N = 2**n
3  m = N/2
```

$$(3.52)$$

For simplicity, we assume in the commands above that the size of the list is $N = 2^n$ and the searched element m is in the middle of the list.

Now, we define the oracle operator and the inversion about the mean. Although Python has native support to complex numbers, they are not necessary here and we can assume that all the matrices have floating point entries.

```
1  Rf = np. zeros ((N,N))
2  Rf[m,m] = 1
3  Rf = −2*Rf+np.eye(N)
4  Rmean = 2*np.ones((N,N), dtype=float )/N − np.eye(N)
```

$$(3.53)$$

We can finally define Grover's operator, simply by the composition of both matrices:

```
1  G = np.dot(Rmean,Rf)
```

$$(3.54)$$

Alternatively, we can define Grover's operator directly from the general form of the entries of its matrix. The following Python subroutine returns the i-th row, and the j-th column of Grover's matrix G:

```
1   def grover_entry (i, j):
2       c1 = 2.0/N
3       if j == m:
4           if i == j:
5               return 1.0−c1
6           else :
7               return −c1
8       else :
9           if i == j:
10              return c1−1.0
11          else :
12              return c1
```

$$(3.55)$$

Whichever option we had chosen in order to define the matrix G in Python, now we have to initialize the state of the quantum computer. We could start at state $|00\ldots0\rangle$,

```
1   psi = np.zeros(N, dtype=complex)
2   psi [0] = 1.0
```

$$(3.56)$$

and then apply a Hadamard transformation in order to generate the uniform superposition. The matrix for Hadamard transformation can be obtained from *NumPy* commands, similarly to the examples above. For simplicity, let us just define the uniform superposition directly, which can be done by a single line in Python:

```
1   psi = np.ones(N, dtype=complex)/np.sqrt (N)
```

$$(3.57)$$

Now we are ready to apply Grover's operator to the quantum state and plot the results:

```
1   psi_t = psi
2   totalSteps = int ((np.pi/4)*np.sqrt (N))
3
4   for t in range( totalSteps ):
5       # evolution
6       psi_t = np.dot(U, psi_t)
7
8       # plotting commands
9       prob = np.abs( psi_t )**2
10      plt . xlabel ('element_index')
11      plt . ylabel (' probability ')
12      plt . ylim ([0.0,1.0])
13      plt . plot (prob)
14      plt . show()
```

$$(3.58)$$

The commands above will plot the probability distribution after each step of the algorithm. Although it may be useful to visualize the evolution of the algorithm, that may generate a massive amount of plots. If the reader is running the experiment with a large input, then it is more convenient to leave the plotting commands outside the *for* loop—just by removing indentation—which will cause the script to plot only the final distribution.

One can also use *SymPy*, a Python library for symbolic mathematics. It includes specific commands for simulations of Grover's algorithm and other quantum algorithms. There is also an implementation of Grover's algorithm for IBM Quantum Experience in [7].

References

1. Aaronson, S.: Limits on Efficient Computation in the Physical World. Ph.D. thesis, University of California, Berkeley (2004). ArXiv: quant-ph/0412143
2. Arora, S., Barak, B.: Computational Complexity: A Modern Approach. Cambridge University Press, Cambridge (2009)
3. Bennett, C.H., Bernstein, E., Brassard, G., Vazirani, U.: Strengths and weaknesses of quantum computing. SIAM J. Comput. **26**(5), 1510–1523 (1997). https://doi.org/10.1137/S0097539796300933
4. Biron, D., Biham, O., Biham, E., Grassl, M., Lidar, D.A.: Generalized grover search algorithm for arbitrary initial amplitude distribution. In: Williams, C.P. (ed.) Quantum Computing and Quantum Communications, pp. 140–147. Springer, Berlin (1999)
5. Boyer, M., Brassard, G., Høyer, P., Tapp, A.: Tight bounds on quantum searching. Fortschritte der Physik **46**(4-5), 493–505 (1998). https://doi.org/10.1002/(SICI)1521-3978(199806)46:4/5<493::AID-PROP493>3.0.CO;2-P
6. Brickman, K.A., Haljan, P.C., Lee, P.J., Acton, M., Deslauriers, L., Monroe, C.: Implementation of grover's quantum search algorithm in a scalable system. Phys. Rev. A **72**, 050, 306 (2005). https://doi.org/10.1103/PhysRevA.72.050306
7. Coles, P.J., Eidenbenz, S., Pakin, S., Adedoyin, A., Ambrosiano, J., Anisimov, P., Casper, W., Chennupati, G., Coffrin, C., Djidjev, H., Gunter, D., Karra, S., Lemons, N., Lin, S., Lokhov, A., Malyzhenkov, A., Mascarenas, D., Mniszewski, S., Nadiga, B., O'Malley, D., Oyen, D., Prasad, L., Roberts, R., Romero, P., Santhi, N., Sinitsyn, N., Swart, P., Vuffray, M., Wendelberger, J., Yoon, B., Zamora, R., Zhu, W.: Quantum Algorithm Implementations for Beginners. Technical Report LA-UR:2018-229, Los Alamos National Laboratory, Los Alamos, New Mexico (2018)
8. Farhi, E., Gutmann, S.: An Analog Analogue of a Digital Quantum Computation. Technical Report MIT-CTP-2593, Massachusetts Institute of Technology (1996). ArXiv:quant-ph/9612026
9. Farhi, E., Goldstone, J., Gutmann, S., Sipser, M.: Quantum Computation by Adiabatic Evolution. Technical Report MIT-CTP-2936, Massachusetts Institute of Technology (2000). ArXiv:quant-ph/0001106
10. Godfrin, C., Ferhat, A., Ballou, R., Klyatskaya, S., Ruben, M., Wernsdorfer, W., Balestro, F.: Operating quantum states in single magnetic molecules: implementation of grover's quantum algorithm. Phys. Rev. Lett. **119**, 187,702 (2017). https://doi.org/10.1103/PhysRevLett.119.187702
11. Grover, L.K.: A fast quantum mechanical algorithm for database search. In: Proceedings of the Twenty-Eighth Annual ACM Symposium on Theory of Computing-STOC '96, pp. 212–219. ACM, New York (1996). https://doi.org/10.1145/237814.237866

12. Grover, L.K.: Quantum computers can search arbitrarily large databases by a single query. Phys. Rev. Lett. **79**(23), 4709–4712 (1997). https://doi.org/10.1103/PhysRevLett.79.4709

13. Grover, L.K.: Quantum mechanics helps in searching for a needle in a haystack. Phys. Rev. Lett. **79**(2), 325–328 (1997). https://doi.org/10.1103/PhysRevLett.79.325

14. Grover, L.K.: Quantum computers can search rapidly by using almost any transformation. Phys. Rev. Lett. **80**(19), 4329–4332 (1998). https://doi.org/10.1103/PhysRevLett.80.4329

15. Jones, J.A., Mosca, M., Hansen, R.H.: Implementation of a quantum search algorithm on a quantum computer. Nature **393**, 344–346 (1998). https://doi.org/10.1038/30687

16. Kowada, L., Lavor, C., Portugal, R., Figueiredo, C.: A new quantum algorithm for solving the minimum searching problem. Int. J. Quantum Inf. **6**, 427–436 (2008)

17. Lara, P., Portugal, R.P., Lavor, C.: A new hybrid classical-quantum algorithm for continuous global optimization problems. J. Glob. Optim. **60**, 317–331 (2014)

18. Lavor, C., Carvalho, L., Portugal, R., Moura, C.: Complexity of grover's algorithm: an algebraic approach. Int. J. Appl. Math. **20**, 801–814 (2007)

19. Leuenberger, M.N., Loss, D.: Quantum computing in molecular magnets. Nature **410**, 789–793 (2001). https://doi.org/10.1038/35071024

20. Long, G.L.: Grover algorithm with zero theoretical failure rate. Phys. Rev. A **64**, 022,307 (2001). https://doi.org/10.1103/PhysRevA.64.022307

21. Mosca, M.: Counting by quantum eigenvalue estimation. Theor. Comput. Sci. **264**(1), 139–153 (2001). https://doi.org/10.1016/S0304-3975(00)00217-6

22. Muthukrishnan, S., Lidar, D.A.: Quasiadiabatic Grover search via the Wentzel-Kramers-Brillouin approximation. Phys. Rev. A **96**, 012,329 (2017). https://doi.org/10.1103/PhysRevA.96.012329

23. Nielsen, M.A., Chuang, I.L.: Quantum Computation and Quantum Information. Cambridge University Press, Cambridge (2010). https://doi.org/10.1017/CBO9780511976667

24. Portugal, R.: Quantum Walks and Search Algorithms. Springer, Cham (2018). https://doi.org/10.1007/978-3-319-97813-0

25. Roland, J., Cerf, N.J.: Quantum search by local adiabatic evolution. Phys. Rev. A **65**, 042,308 (2002). https://doi.org/10.1103/PhysRevA.65.042308

26. Roland, J., Cerf, N.J.: Quantum-circuit model of hamiltonian search algorithms. Phys. Rev. A **68**, 062,311 (2003). https://doi.org/10.1103/PhysRevA.68.062311

27. Vandersypen, L.M.K., Steffen, M., Sherwood, M.H., Yannoni, C.S., Breyta, G., Chuang, I.L.: Implementation of a three-quantum-bit search algorithm. Appl. Phys. Lett. **76**(5), 646–648 (2000). https://doi.org/10.1063/1.125846

28. Zalka, C.: Grover's quantum searching algorithm is optimal. Phys. Rev. A **60**(4), 2746–2751 (1999). https://doi.org/10.1103/PhysRevA.60.2746

29. Zhang, W., Cheng, K., Wu, C., Wang, Y., Li, H., Zhang, X.: Implementing quantum search algorithm with metamaterials. Adv. Mater. **30**(1), 1703986 (2018). https://doi.org/10.1002/adma.201703986.

Chapter 4
Shor's Algorithm for Integer Factorization

The general number field sieve is the most efficient classical algorithm known for factorizing integers larger than 10^{100}. The second fastest method for integer factorization is the quadratic sieve, which is most recommended for factorizing numbers smaller than 10^{100}. Other important classical algorithms are the Lenstra elliptic curve method [14] and the Pollard's rho method [20]. The time complexity of each of those algorithms is exponential in the number of bits of the input. The time complexity of the general field sieve algorithm, for instance, is of the form

$$\exp\left(\left(\sqrt[3]{\frac{64}{9}} + o(1)\right)(\ln N)^{1/3}(\ln \ln N)^{2/3}\right), \tag{4.1}$$

where N is the number to be factorized. In fact, the difficulty of factorizing large integers on classical computers is at the heart of many cryptography protocols widely used nowadays.[1] Perhaps that is the reason why Shor's algorithm is still one of the most celebrated breakthroughs in quantum algorithm design. The original presentations of Shor's algorithm are references [21] and [22]. Shor's algorithm finds the prime factors of a composite number N in polynomial time on the number of input bits. A key ingredient of Shor's algorithm is the Quantum Fourier Transform, which has an exponential speed-up over the classical Fast Fourier Transform [10].

---

[1]There are several good references on the subject of integer factorization and classical cryptography. The reader may refer to Hoffstein, Pipher and Silverman [9], or to Crandall and Pomerance [5], or to Wagstaff [24], for instance.

© The Author(s), under exclusive license to Springer Nature Switzerland AG 2019
F. de Lima Marquezino et al., *A Primer on Quantum Computing*, SpringerBriefs in
Computer Science, https://doi.org/10.1007/978-3-030-19066-8_4

4.1   A Reduction of Integer Factorization to Order Finding

Think of a large number such as one with 300 digits in decimal notation. Though N is large, the number of qubits necessary to store it is small. In general, $\log_2 N$ is not an integer, so let us define

$$n = \lceil \log_2 N \rceil. \tag{4.2}$$

A quantum computer with n qubits can store N or any other positive integer less than N. With a little thought, we see that the number of prime factors of N is at most n. If both the number of qubits and the number of factors are less than or equal to n, then it is natural to ask if there is a quantum algorithm that factors N in a number of steps which is polynomial in n. In fact, this question has a positive answer, and in order to describe it, we first need to understand how the problem of factorizing an integer N can be efficiently reduced to the problem of finding the *order* of a number.

We should start by choosing a random integer x less than N. We can efficiently[2] find the greatest common divisor—or GCD, for short—between x and N. If x and N have common factors, then GCD(x, N) gives a factor of N, and in this lucky case the problem is solved! Therefore, it suffices to investigate the case when x is coprime to N, which is much harder.

The order of x modulo N is defined as the least positive integer r such that

$$x^r \equiv 1 \mod N. \tag{4.3}$$

The notation $a \equiv b \mod N$ means that both a and b yields the same remainder when divided by N. If r is even, we can define y by

$$y \equiv x^{r/2} \mod N, \tag{4.4}$$

with $0 \le y < N$. Since r depends on x, if r is not even, we have to start the process all over again by randomly choosing another x. Notice that y satisfies $y^2 \equiv 1 \mod N$, or equivalently $(y-1)(y+1) \equiv 0 \mod N$, which means that N divides $(y-1)(y+1)$. If $1 < y < N-1$, the factors $y-1$ and $y+1$ satisfy $0 < y-1 < y+1 < N$, therefore N cannot divide $y-1$ nor $y+1$ separately. The only alternative is that both $y-1$ and $y+1$ have factors of N, and they yield N by multiplication. Hence, GCD$(y-1, N)$ and GCD$(y+1, N)$ yield non trivial factors of N. If N still has remaining factors, they can be calculated by applying the algorithm recursively.

Consider $N = 21$ as an example. The sequence of equivalences

---

[2]By using Euclid's algorithm, for instance. The ancient greek mathematician Euclid described this algorithm in his *Elements*, c. 300 BC.

$$2^4 \equiv 16 \mod 21$$

$$2^5 \equiv 11 \mod 21 \qquad (4.5)$$

$$2^6 \equiv 1 \mod 21$$

show that the order of two modulo 21 is $r = 6$. Therefore, $y \equiv 2^3 \equiv 8 \mod 21$. The prime factors of 21 are given by $y - 1$, which yields the factor seven, and by $y + 1$, which yields the factor three. For a more detailed study of number theory, the reader may look for Gathen and Gerhard's book [8].

In summary, if we pick up at random a positive integer x less than N and calculate $GCD(x, N)$, either we have a factor of N or we learn that x is co-prime to N. In the latter case, if x satisfies the conditions (1) that its order r is even, and (2) that $0 < y - 1 < y + 1 < N$, then $GCD(y - 1, N)$ and $GCD(y + 1, N)$ yield factors of N. If one of the conditions is not true, we start over until finding a proper candidate x. The method would not be useful if these assumptions were too restrictive, but fortunately that is not the case. The method systematically fails if N is a power of some odd prime, but an alternative efficient classical algorithm for this case is known. If N is even, we can keep dividing by two until the result turns out to be odd. It remains to apply the method for odd composite integers that are not a power of some prime number. It is cumbersome to prove that the probability of finding x coprime to N satisfying the conditions (1) and (2) is high—in fact, this probability is $1 - 1/2^{k-1}$, where k is the number of prime factors of N.[3] The worst case is when N has only two prime factors, then the probability is greater than or equal to $1/2$.

At first sight, it seems that we have just described an efficient algorithm to find a factor of N. That is not yet true, since it is not known an efficient *classical* algorithm to calculate the order of an integer x modulo N. On the other hand, there is an efficient *quantum* algorithm, which is precisely Shor's algorithm.[4] Let us describe it.

4.2   Quantum Algorithm to Calculate the Order

Consider the circuit of Fig. 4.1. It calculates the order r of the positive integer x less than N, coprime to N. The subroutine V_x is given by the unitary linear operator

$$V_x \, |j\rangle \, |k\rangle = |j\rangle \, \Big| k \oplus (x^j \mod N)\Big), \qquad (4.6)$$

---

[3] The proof can be found in Appendix B of reference [7].
[4] The problem of finding order can be generalized as the hidden subgroup problem [1, 11]. Several quantum algorithms can be described under the same framework of the hidden subgroup problem [2, 17].

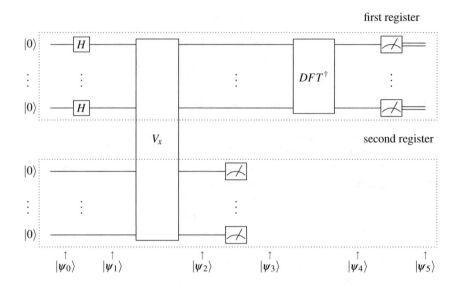

Fig. 4.1 Quantum circuit for finding the order of the positive integer x modulo N

where $|j\rangle$ and $|k\rangle$ are the states of the first and second registers, respectively, and $\oplus$ is the bitwise binary sum.[5] The subroutine DFT is the Discrete Fourier Transform operator, which will be described ahead.

The first register has t qubits, where t is generally chosen such that $N^2 \leq 2^t < 2N^2$, for reasons that will become clear later on. As an exception, if the order r is a power of 2, then it is enough to take $t = \log_2 N = n$. In this section, we consider this very special case and leave the general case for Sect. 4.4. We will keep the variable t in order to generalize the discussion later on. The second register has n qubits.

It is interesting to notice that, on a very important experiment with nuclear magnetic resonance, Vandersypen et al. bypassed part of Shor's algorithm and managed to factorize $N = 15$ on a quantum computer with only seven qubits [23]. A more recent implementation of Shor's algorithm performed by Monz et al. has the advantage of being scalable [18]. A key ingredient in that result was Kitaev's version of the Quantum Fourier Transform [12].

The states of the quantum computer are indicated by the vectors $|\psi_0\rangle$ to $|\psi_5\rangle$ in Fig. 4.1. The initial state is

$$|\psi_0\rangle = \underbrace{|00\ldots0\rangle}_{t \text{ times}} \underbrace{|00\ldots0\rangle}_{n \text{ times}}. \tag{4.7}$$

---

[5]At this point, k and x^j modulo N must be converted into base-2 notation in order to execute the binary sum.

Recall that the application of the Hadamard operator

$$H = \frac{1}{\sqrt{2}} \begin{bmatrix} 1 & 1 \\ 1 & -1 \end{bmatrix} \tag{4.8}$$

on each qubit of the first register yields

$$|\psi_1\rangle = \frac{1}{\sqrt{2^t}} \sum_{j=0}^{2^t-1} |j\rangle |0\rangle . \tag{4.9}$$

The first register is then in a superposition of all states of the computational basis with equal amplitudes given by $1/\sqrt{2^t}$. Now, notice that when we apply V_x to $|\psi_1\rangle$, we yield the state

$$|\psi_2\rangle = \frac{1}{\sqrt{2^t}} \sum_{j=0}^{2^t-1} V_x |j\rangle |0\rangle$$

$$= \frac{1}{\sqrt{2^t}} \sum_{j=0}^{2^t-1} |j\rangle \left|x^j\right\rangle . \tag{4.10}$$

The state $|\psi_2\rangle$ is remarkable. Because V_x is linear, it acts on all basis states $|j\rangle |0\rangle$ for 2^t values of j, so this generates all powers of x simultaneously. This feature is the *quantum parallelism*, the same property observed in Grover's algorithm. Some of these powers are equal to one, which correspond to the states $|0\rangle |1\rangle , |r\rangle |1\rangle , |2r\rangle |1\rangle , \ldots , \left|\left(\frac{2^t}{r} - 1\right)r\right\rangle |1\rangle$. This explains the choice for V_x given by Eq. (4.6). Classically, one would calculate successively x^j, for j starting from two until reaching $j = r$. Quantumly, on the other hand, one can calculate all powers of x with just one application of V_x. At the quantum level, the values of j that yield $x^j \equiv 1 \mod N$ are "known". However, this quantum information is not fully available at the classical level. A classical information of a quantum state is obtained by practical measurements and, at this point, it does not help if we measure the first register, since all states in the superposition given by Eq. (4.10) have equal amplitudes. The first part of the strategy to find r is to observe that the first register of the states $|0\rangle |1\rangle , |r\rangle |1\rangle , |2r\rangle |1\rangle , \ldots , \left|2^t - r\right\rangle |1\rangle$ is periodic. Hence, the information we want is a period!

In order to simplify the calculation, let us measure the second register. Before doing this, we will rewrite $|\psi_2\rangle$ collecting equal terms in the second register. Since x^j is a periodic function with period r, substitute j by $ar + b$ in Eq. (4.10), where $0 \le a \le (2^t/r) - 1$ and $0 \le b \le r - 1$. Recall that we are supposing that $t = n$ and that r is a power of 2, therefore r divides 2^t. Thus, Eq. (4.10) is converted to

$$|\psi_2\rangle = \frac{1}{\sqrt{2^t}} \sum_{b=0}^{r-1} \left(\sum_{a=0}^{\frac{2^t}{r}-1} |ar+b\rangle \right) |x^b\rangle. \tag{4.11}$$

In the second register, we have replaced x^{ar+b} by x^b, since $x^r \equiv 1 \mod N$. Now, the second register is measured.[6] Any output $x^0, x^1, \ldots, x^{r-1}$ can be obtained with equal probability. Suppose that the result is x^{b_0}. The state of the quantum computer is now

$$|\psi_3\rangle = \sqrt{\frac{r}{2^t}} \sum_{a=0}^{\frac{2^t}{r}-1} |ar+b_0\rangle |x^{b_0}\rangle. \tag{4.12}$$

Notice that after the measurement, the constant is renormalized to $\sqrt{r/2^t}$, since there are $2^t/r$ terms[7] in the sum in Eq. (4.12). Figure 4.2 shows the probability of obtaining the states of the computational basis upon measuring the first register. The probabilities form a periodic function with period r. Their values are zero except for the states $|b_0\rangle$, $|r+b_0\rangle$, $|2r+b_0\rangle$, ..., $|2^t - r + b_0\rangle$.

How can one find out the period of a function efficiently? The answer is in the Fourier transform. The Fourier transform of a periodic function with period r is a new periodic function with period proportional to $1/r$. This makes a difference

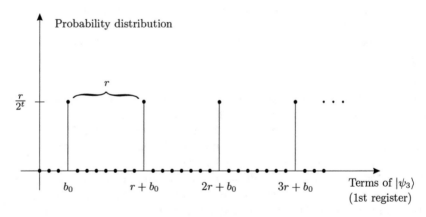

Fig. 4.2 Probability distribution of $|\psi_3\rangle$ measured in the computational basis (for the case $b_0 = 3$ and $r = 8$). The horizontal axis has 2^t points. The number of peaks is $2^t/r$ and the period is r

---

[6]In Nielsen and Chuang's book, there is a proof that measurements can always be moved from an intermediate stage of a quantum circuit to the end of the circuit without changing the results [19], so the intermediate measurement in Shor's algorithm is not really necessary—although it simplifies the remaining calculations.

[7]Or equivalently, there are r terms inside the parenthesis in Eq. (4.11).

for finding r. The Fourier transform is the second and last part of the strategy. The whole method relies on an efficient quantum algorithm for calculating the Fourier transform, which is not available classically. In Sect. 4.5, we show that the Fourier transform is calculated efficiently in a quantum computer.

4.3   The Quantum Discrete Fourier Transform

The classical algorithm of Fast Fourier Transform was developed by Cooley and Tukey [4] and an excellent exposition of it can be found in Donald Knuth's famous *The Art of Computer Programming* [13]. The quantum algorithm for Discrete Fourier Transform can be derived from the classical algorithm, as described in Ref. [16].

The Fourier transform of the function $F : \{0, \ldots, N-1\} \to \mathbb{C}$ is a new function $\tilde{F} : \{0, \ldots, N-1\} \to \mathbb{C}$ defined as

$$\tilde{F}(k) = \frac{1}{\sqrt{N}} \sum_{j=0}^{N-1} e^{2\pi i j k/N} F(j). \tag{4.13}$$

We can apply the Fourier transform either on a function or on the states of the computational basis. The Fourier transform applied to the state $|k\rangle$ of the computational basis $\{|0\rangle, \ldots, |N-1\rangle\}$ is

$$\text{DFT} |k\rangle = |\psi_k\rangle$$

$$= \frac{1}{\sqrt{N}} \sum_{j=0}^{N-1} e^{2\pi i j k/N} |j\rangle, \tag{4.14}$$

where the set $\{|\psi_k\rangle : k = 0, \ldots, N-1\}$ forms a new orthonormal basis. The Fourier transform is a unitary linear operator. Hence, if we know how it acts on the states of the computational basis, we also know how it acts on an arbitrary state

$$|\psi\rangle = \sum_{a=0}^{N-1} F(a) |a\rangle. \tag{4.15}$$

The Fourier transform of $|\psi\rangle$ can be performed indistinctly using either Eq. (4.13) or (4.14). We will use the latter.

To prove that $\{|\psi_k\rangle : k = 0, \ldots, N-1\}$ is an orthonormal basis, i.e.,

$$\langle \psi_{k'} | \psi_k \rangle = \delta_{k'k}, \tag{4.16}$$

Fig. 4.3 Vectors $e^{2\pi ij/7}$, for $j = 0, \ldots, 6$, in the complex plane. Their sum is zero by symmetry arguments. This is an example of Eq. (4.17) for $N = 7, k = 1$

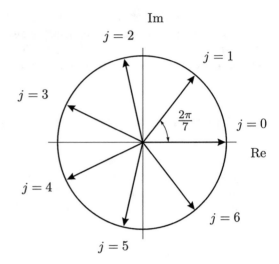

we can use the identity

$$\frac{1}{N} \sum_{j=0}^{N-1} e^{2\pi ijk/N} = \begin{cases} 1, & \text{if } k \text{ is a multiple of } N \\ 0, & \text{otherwise,} \end{cases} \tag{4.17}$$

which is useful in the context of Fourier transforms. It is not difficult to prove that Eq. (4.17) is true. If k is a multiple of N, then $e^{2\pi ijk/N} = 1$ and the first case of the identity follows. If k is not a multiple of N, Eq. (4.17) is true even if N is not a power of two. In Fig. 4.3, we have each term $e^{2\pi ijk/N}$, for the case where $k = 1$ and $N = 7$, as vectors in the complex plane. Note that the sum of vectors must be zero by a symmetry argument: the distribution of vectors is isotropic. Usually it is said that the interference is destructive in this case.

Using this identity, we can define the inverse Fourier transform, which is similar to Eq. (4.14), just with a minus sign on the exponent. Note that $\text{DFT}^{-1} = \text{DFT}^{\dagger}$, since DFT is a unitary operator.

We will present the details of a quantum circuit to perform the Fourier transform in Sect. 4.5. Now we will continue the calculation process of the circuit of Fig. 4.1. We are ready to find out $|\psi_4\rangle$, the next state of the quantum computer. Applying the inverse Fourier transform on the first register, using Eq. (4.14) and the linearity of $\text{DFT}^{\dagger}$, we obtain

$$|\psi_4\rangle = \text{DFT}^{\dagger} |\psi_3\rangle$$

$$= \sqrt{\frac{r}{2^t}} \sum_{a=0}^{\frac{2^t}{r}-1} \left(\frac{1}{\sqrt{2^t}} \sum_{j=0}^{2^t-1} e^{-2\pi ij(ar+b_0)/2^t} |j\rangle \right) |x^{b_0}\rangle. \tag{4.18}$$

Inverting the summation order, we have

$$|\psi_4\rangle = \frac{1}{\sqrt{r}} \left(\sum_{j=0}^{2^t-1} \left[\frac{1}{2^t/r} \sum_{a=0}^{\frac{2^t}{r}-1} e^{\frac{-2\pi i j a}{2^t/r}} \right] e^{-2\pi i j b_0/2^t} |j\rangle \right) \left| x^{b_0} \right\rangle. \qquad (4.19)$$

Using Eq. (4.17), we see that the expression in square brackets is zero except when $j = k2^t/r$, with $k = 0, \ldots, r - 1$. When j takes such values, the expression in the square brackets is equal to one. Thus, we have

$$|\psi_4\rangle = \frac{1}{\sqrt{r}} \left(\sum_{k=0}^{r-1} e^{-2\pi i \frac{k}{r} b_0} \left| \frac{k2^t}{r} \right\rangle \right) \left| x^{b_0} \right\rangle. \qquad (4.20)$$

In order to find r, the expression for $|\psi_4\rangle$ in Eq. (4.20), has two advantages over the expression for $|\psi_3\rangle$, in Eq. (4.12). The first advantage is that r is in the denominator of the ket label, and the second advantage is that the random parameter b_0 moved from the ket label to the exponent occupying now a harmless place. In Fig. 4.4, we have the probability distribution of $|\psi_4\rangle$ measured in the computational basis. Measuring the first register, we get the value $k_0 2^t/r$, where k_0 can be any number between 0 and $r - 1$ with equal probability—see the peaks in Fig. 4.4. If we obtain $k_0 = 0$, we have no clue at all about r, and the algorithm must be run again. If $k_0 \neq 0$, we divide $k_0 2^t/r$ by 2^t, obtaining k_0/r. Neither k_0 nor r are known. If k_0 is coprime to r, we simply select the denominator.

If k_0 and r have a common factor, the denominator of the reduced fraction k_0/r is a factor of r but not r itself. Suppose that the denominator is r_1. Let $r = r_1 r_2$. Now the goal is to find r_2, which is the order of x^{r_1}. We run again the quantum part of the algorithm to find the order of x^{r_1}. If we find r_2 this time, the algorithm

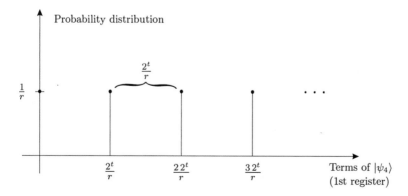

Fig. 4.4 Probability distribution of $|\psi_4\rangle$ measured in the computational basis. The horizontal axis has 2^t points, only the non-null terms are shown. The number of peaks is r and the period is $2^t/r$

halts, otherwise we keep applying it recursively. The recursive process is efficient, because the number of iterations is less than or equal to $\log_2 r$.

Take $N = 15$ as an example, which is the least nontrivial composite number. The set of numbers less than 15 that are coprime to 15 is $\{1, 2, 4, 7, 8, 11, 13, 14\}$. The numbers in the set $\{4, 11, 14\}$ have order two and the numbers in the set $\{2, 7, 8, 13\}$ have order four. Therefore, in any case r is a power of two and the factors of $N = 15$ can be found in a 8-bit quantum computer—because $t + n = 2\lceil \log_2 15 \rceil = 8$.

4.4 Generalization by Means of an Example

In the previous sections, we have considered a special case when the order r is a power of two and $t = n$ (t is the number of qubits in the first register—see Fig. 4.1— and $n = \lceil \log_2 N \rceil$). In this section, we consider the factorization of $N = 21$, that is the next nontrivial composite number. We must choose t such that 2^t is between N^2 and $2N^2$, which is always possible. For $N = 21$, the smallest value of t is nine. This is the simplest example allowed by the constraints, but enough to display all properties of Shor's algorithm.

The first step is to pick up x at random such that $1 < x < N$, and to test whether x is coprime to N. If not, we easily find a factor of N by calculating $GCD(x, N)$. If yes, the quantum part of the algorithm starts. Suppose that $x = 2$ has been chosen. The goal is to find out that the order of x is $r = 6$. The quantum computer is initialized in the state

$$|\psi_0\rangle = |0\rangle |0\rangle , \tag{4.21}$$

where the first register has $t = 9$ qubits and the second has $n = 5$ qubits. Next step is the application of $H^{\otimes 9}$ on the first register yielding

$$|\psi_1\rangle = \frac{1}{\sqrt{512}} \sum_{j=0}^{511} |j\rangle |0\rangle , \tag{4.22}$$

as in Eq. (4.9). The next step is the application of V_x as defined in Eq. (4.6), which yields

$$|\psi_2\rangle = \frac{1}{\sqrt{512}} \sum_{j=0}^{511} |j\rangle \left| 2^j \bmod N \right\rangle$$

$$= \frac{1}{\sqrt{512}} \Big(|0\rangle |1\rangle + |1\rangle |2\rangle + |2\rangle |4\rangle + |3\rangle |8\rangle + |4\rangle |16\rangle + |5\rangle |11\rangle +$$

$$|6\rangle |1\rangle + |7\rangle |2\rangle + |8\rangle |4\rangle + |9\rangle |8\rangle + |10\rangle |16\rangle + |11\rangle |11\rangle +$$

$$|12\rangle |1\rangle + \dots \Big). \tag{4.23}$$

Notice that the above expression has the following pattern: the states of the second register of each "column" are the same. Therefore we can rearrange the terms in order to collect the second register:

$$|\psi_2\rangle = \frac{1}{\sqrt{512}}\Big[\big(\,|0\rangle + |6\rangle + |12\rangle + \ldots + |504\rangle + |510\rangle\,\big)\,|1\rangle +$$

$$\big(\,|1\rangle + |7\rangle + |13\rangle + \ldots + |505\rangle + |511\rangle\,\big)\,|2\rangle +$$

$$\big(\,|2\rangle + |8\rangle + |14\rangle + \ldots + |506\rangle\,\big)\,|4\rangle +$$

$$\big(\,|3\rangle + |9\rangle + |15\rangle + \ldots + |507\rangle\,\big)\,|8\rangle +$$

$$\big(\,|4\rangle + |10\rangle + |16\rangle + \ldots + |508\rangle\,\big)\,|16\rangle +$$

$$\big(\,|5\rangle + |11\rangle + |17\rangle + \ldots + |509\rangle\,\big)\,|11\rangle\Big]. \tag{4.24}$$

This feature was made explicit in Eq. (4.11). Since the order is not a power of two, there is a small difference here: the first two lines of Eq. (4.24) have 86 terms, while the remaining ones have 85.

Now one measures the second register, yielding one of the numbers from set $\{1, 2, 4, 8, 16, 11\}$ equiprobably. Suppose that the result of the measurement is two, then

$$|\psi_3\rangle = \frac{1}{\sqrt{86}}\,(|1\rangle + |7\rangle + |13\rangle + \ldots + |505\rangle + |511\rangle)\,|2\rangle . \tag{4.25}$$

Notice that the state $|\psi_3\rangle$ was renormalized in order to have unit norm. It does not matter what is the result of the measurement; what matters is the periodic pattern of Eq. (4.25). The period of the states of the first register is the solution to the problem and the Fourier transform can reveal the value of the period. Thus, the next step is the application of the inverse Fourier transform on the first register of $|\psi_3\rangle$, yielding

$$|\psi_4\rangle = \mathrm{DFT}^\dagger\,|\psi_3\rangle$$

$$= \mathrm{DFT}^\dagger\left(\frac{1}{\sqrt{86}}\sum_{a=0}^{85}|6a + 1\rangle\right)|2\rangle$$

$$= \frac{1}{\sqrt{512}}\sum_{j=0}^{511}\left(\left[\frac{1}{\sqrt{86}}\sum_{a=0}^{85}e^{-2\pi i\frac{6ja}{512}}\right]e^{-2\pi i\frac{j}{512}}\,|j\rangle\right)|2\rangle , \tag{4.26}$$

where we have used Eq. (4.14) and have rearranged the sums. The last equation is similar to Eq. (4.19), although with an important difference. In Sect. 4.2, we were assuming that r divides 2^t. This is not true in the present example, since 6 does not divide 512, and thus we cannot use the identity of Eq. (4.17) to simplify the term in brackets in Eq. (4.26). This term never vanishes, but its main contribution is

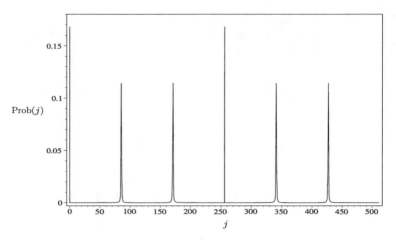

Fig. 4.5 Plot of $\text{Prob}(j)$ against j. Compare to the plot of Fig. 4.4, where peaks are not spread and have the same height

still around $j = 0, 85, 171, 256, 341, 427$, which are obtained rounding $512k_0/6$ for k_0 from 0 to 5—compare to the discussion that follows Eq. (4.20). In order to observe this, let us plot the probability of getting the result j, in the interval 0 to 511, by measuring the first register of state $|\psi_4\rangle$. From Eq. (4.26), we have that the probability is

$$\text{Prob}(j) = \frac{1}{512 \times 86} \left| \sum_{a=0}^{85} e^{-2\pi i \frac{6ja}{512}} \right|^2. \tag{4.27}$$

The plot of $\text{Prob}(j)$ is shown in Fig. 4.5. We see the peaks around $j = 0, 85, 171, 256, 341, 427$, indicating a high probability of getting one of these values, or some value very close to them. In between, the probability is almost zero. The sharpness of the peaks depends on t—the number of qubits in the first register. The lower limit $2^t \geq N^2$ ensures a high probability in measuring a value of j carrying the desired information. Lomonaco performed a careful analysis of the expression in Eq. (4.27) [15], and Einarson performed a meticulous study of the peak form of Fig. 4.5 [6].

Let us analyze the possible measurement results. If we get $j = 0$, the first peak, the algorithm has failed in this round. It must be run again. We keep $x = 2$ and rerun the quantum part of the algorithm. From Eq. (4.27), we have that $\text{Prob}(0) = 86/512 \approx 0.167$, and thus the probability of getting $j = 0$ is low.

Now suppose we get $j = 85$ or any value in the second peak. We divide the result by 512 yielding $85/512$, which is a rational approximation of $k_0/6$, for $k_0 = 1$. How can we obtain r from $85/512$? The method of continued fraction approximation allows one to extract the desired information. A general continued fraction expansion of a rational number j_1/j_2 has the form

$$\frac{j_1}{j_2} = a_0 + \cfrac{1}{a_1 + \cfrac{1}{...+\frac{1}{a_p}}}, \tag{4.28}$$

usually represented as $[a_0, a_1, \ldots, a_p]$, where a_0 is a non-negative integer and $a_1, \ldots, a_p$ are positive integers. The q-th convergent, $0 \leq q \leq p$, is defined as the rational number $[a_0, a_1, \ldots, a_q]$. It is an approximation to j_1/j_2 and has a denominator smaller than j_2.

This method is easily applied by inversion of the fraction followed by integer division with rational remainder. Inverting 85/512 yields 512/85, which is equal to $6 + 2/85$. We repeat the process with 2/85 until we get the numerator equal to one. The result is

$$\frac{85}{512} = \cfrac{1}{6 + \cfrac{1}{42 + \frac{1}{2}}}. \tag{4.29}$$

Therefore, the convergents of 85/512 are 1/6, 42/253, and 85/512. We must select the convergents that have a denominator smaller than $N = 21$, since $r < N$. The inequality $r \leq \varphi(N)$ follows from the Euler's theorem, which states that $x^{\varphi(N)} \equiv 1$ mod N, where x is a positive integer coprime to N and φ is the Euler's totient function, which gives the number of positive integers less than N, coprime to N. The inequality $\varphi(N) < N$ follows from the definition of φ.

Returning to our example, we find then that the method of continued fraction approximation yields 1/6, and then $r = 6$. We check that $2^6 \equiv 1 \mod 21$, and then the quantum part of the algorithm halts with the correct answer. The order $r = 6$ is an even number, therefore GCD($2^{(6/2)} \pm 1, 21$) gives two non trivial factors of 21. A straightforward calculation shows that any measured result in the second peak, say, $81 \leq j \leq 89$, yields the convergent 1/6.

Consider now the third peak, which corresponds to $k_0/6$, for $k_0 = 2$. We apply again the method of continued fraction approximation, which yields 1/3 for any j in the third peak, say, $167 \leq j \leq 175$. In this case, we have obtained $r_1 = 3$, which is only a factor of r, since $2^3 \equiv 8 \neq 1 \mod 21$. We run the quantum part of the algorithm again to find the order of eight. We eventually obtain $r_2 = 2$, which yields $r = r_1 r_2 = 6$. The fourth and fifth peaks yield also factors of r. The last peak is similar to the second, yielding r directly.

The general account of the succeeding probability is as follows. The area under each peak is approximately 0.167. The first and fourth peaks are very different from the others—they are not spread. To calculate their contribution to the total probability, we take the basis equal to one. The area under the second, third, fifth, and last peaks are calculated by adding up Prob(j), for j running around the center of each peak. Hence, in approximately 17% cases, corresponding to first peak, the algorithm fails. In approximately 33% cases, corresponding to second and sixth peaks, the algorithm returns r in the first round. In approximately 50% cases, corresponding to all other peaks, the algorithm returns r in the second round or more. Now we should calculate the probability of finding r in the second round. For the third and fifth peaks, the remaining factor is $r_2 = 2$. The graph equivalent to Fig. 4.5 in this case has two peaks, then the algorithm returns r_2 in 50% cases. For the fourth peak, the remaining factor is $r = 3$ and the algorithm returns r_2 in 66.6% cases. This amounts to $(2 \times 50\% + 66.6\%)/3$ of 50%, which is equal to around 22%. In summary, the success probability for $x = 2$ is around 55%.

Exercise 4.1 Calculate each step, classical and quantum, for factorizing the number $N = 35$ using Shor's algorithm. Take $x = 2$ as the required random integer. If that "random" integer does not work in the algorithm, take $x = 3, 4, 5$, and so on.

Exercise 4.2 Give a circuit for V_x explicitly in terms of universal quantum gates. Consider that the circuit would be used for factorizing the number $N = 15$, and take $x = 2$.

Exercise 4.3 Consider that you want to perform the Fourier transform of an arbitrary basis vector of four qubits, say, $|j_1 j_2 j_3 j_4\rangle$, on the computational basis over four qubits.

(a) Show the full quantum circuit for it in terms of universal gates.
(b) Show the matrix representation corresponding to the circuit.
(c) Calculate each intermediate step of the algorithm for the arbitrary input.

4.5   Fourier Transform in Terms of the Universal Gates

In the previous section, we have shown that Shor's algorithm is an efficient probabilistic algorithm, assuming that the Fourier transform could be implemented efficiently. In this section, we decompose the Fourier transform in terms of a set of universal gates composed by CNOT and one-qubit gates. This decomposition allows one to measure the efficiency of the quantum discrete Fourier transform and shows how to implement it in an actual quantum computer.

Recall that the Fourier transform of the states of the computational basis is given by

$$\text{DFT}\,|j\rangle = \frac{1}{\sqrt{N}} \sum_{k=0}^{N-1} e^{2\pi i j k/N}\,|k\rangle. \tag{4.30}$$

After observing that the right hand side of Eq. (4.30) has N terms and the computational basis has N states, we may conclude that the time complexity to calculate classically the Fourier transform of the computational basis using Eq. (4.30) is $O(N^2)$, or equivalently $O(2^{2n})$, which means a double exponential growth if we consider the number of input bits. A very important result in computer science was the development of the classical algorithm known as Fast Fourier Transform— usually abbreviated to FFT—which reduced that complexity to $O(n2^n)$. In this context, we show the improvement by recognizing that the right-hand side of Eq. (4.30) is a very special kind of expansion, which can be fully factored. For example, the Fourier transform of the basis $\{|0\rangle, |1\rangle, |2\rangle, |3\rangle\}$ can be written as

$$\text{DFT} \,|0\rangle = \left(\frac{|0\rangle + |1\rangle}{\sqrt{2}}\right) \otimes \left(\frac{|0\rangle + |1\rangle}{\sqrt{2}}\right)$$

$$\text{DFT} \,|1\rangle = \left(\frac{|0\rangle - |1\rangle}{\sqrt{2}}\right) \otimes \left(\frac{|0\rangle + i\,|1\rangle}{\sqrt{2}}\right)$$

$$\text{DFT} \,|2\rangle = \left(\frac{|0\rangle + |1\rangle}{\sqrt{2}}\right) \otimes \left(\frac{|0\rangle - |1\rangle}{\sqrt{2}}\right)$$

$$\text{DFT} \,|3\rangle = \left(\frac{|0\rangle - |1\rangle}{\sqrt{2}}\right) \otimes \left(\frac{|0\rangle - i\,|1\rangle}{\sqrt{2}}\right). \tag{4.31}$$

Note that, in the example of Eq. (4.31), we are using the basis of a two-qubit system in order to factor the right-hand side. Let us now factor the general expression. The first step is to write Eq. (4.30) in the form

$$\text{DFT} \,|j\rangle = \frac{1}{\sqrt{2^n}} \sum_{k_1=0}^{1} \cdots \sum_{k_n=0}^{1} e^{2\pi i j \sum_{l=1}^{n} \frac{k_l}{2^l}} \,|k_1\rangle \otimes \ldots \otimes |k_n\rangle, \tag{4.32}$$

where the ket $|k\rangle$ was converted to the binary base and we have used the expansion $k = \sum_{l=1}^{n} k_l 2^{n-l}$ in the exponent. Using the fact that the exponential of a sum is a product of exponentials, Eq. (4.32) turns into a (non-commutative) product of the following kets:

$$\text{DFT} \,|j\rangle = \frac{1}{\sqrt{2^n}} \sum_{k_1=0}^{1} \cdots \sum_{k_n=0}^{1} \prod_{l=1}^{n} \left(e^{2\pi i j \frac{k_l}{2^l}} \,|k_l\rangle \right). \tag{4.33}$$

Now we factor Eq. (4.33) by interchanging the sums and the product,

$$\text{DFT} \,|j\rangle = \frac{1}{\sqrt{2^n}} \prod_{l=1}^{n} \sum_{k_l=0}^{1} \left(e^{2\pi i j \frac{k_l}{2^l}} \,|k_l\rangle \right). \tag{4.34}$$

We can easily check that the last equation is correct by performing the calculations backwards. Simply expand the product in Eq. (4.34) and then put all sums at the beginning of the resulting expression to obtain Eq. (4.33). By expanding the sum of Eq. (4.34) and then the product, we finally get

$$\text{DFT} \,|j\rangle = \frac{1}{\sqrt{2^n}} \prod_{l=1}^{n} \left(|0\rangle + e^{2\pi i j/2^l} \,|1\rangle \right)$$

$$= \left(\frac{|0\rangle + e^{2\pi i \frac{j}{2}} \,|1\rangle}{\sqrt{2}}\right) \otimes \left(\frac{|0\rangle + e^{2\pi i \frac{j}{2^2}} \,|1\rangle}{\sqrt{2}}\right) \otimes \ldots \otimes \left(\frac{|0\rangle + e^{2\pi i \frac{j}{2^n}} \,|1\rangle}{\sqrt{2}}\right).$$

$$\tag{4.35}$$

The complexity to calculate Eq. (4.35) for one $|j\rangle$ is $O(n)$, since there are n terms in the product and each term can be calculated in constant time. The complexity in the classical calculation of the fast Fourier transform of the whole computational basis is still exponential, $O(n2^n)$, since the calculation is performed on each of the 2^n basis elements, one at a time. On the other hand, the quantum computer uses quantum parallelism, and the Fourier transform of the state

$$|\psi\rangle = \sum_{a=0}^{2^n-1} F(a)\,|a\rangle, \tag{4.36}$$

which has an exponential number of terms, is calculated with one application of the quantum Fourier transform. The Fourier transform of the 2^n basis elements is performed simultaneously, so the complexity of the quantum Fourier transform is measured by the size of its circuit. We now show that it requires $O(n^2)$ gates.

Consider the circuit of Fig. 4.6, which depicts part of the quantum algorithm for Fourier transform. It is easy to check that the value of the qubits $|j_m\rangle$, $m \neq l$, does not change. We should still check what happens with qubit $|j_l\rangle$.

The unitary matrices R_k are defined as

$$R_k = \begin{bmatrix} 1 & 0 \\ 0 & \exp\left(2\pi i \frac{1}{2^k}\right) \end{bmatrix}. \tag{4.37}$$

Each R_k gate in Fig. 4.6 is controlled by qubit $|j_{k+l-1}\rangle$. Therefore, if $j_{k+l-1} = 0$, then R_k must be replaced by the identity matrix, which means that no action will be performed, and if $j_{k+l-1} = 1$, then R_k must be applied to qubit $|j_l\rangle$. This means that, for calculation purposes, gates R_k controlled by qubit $|j_{k+l-1}\rangle$ can be replaced by "one-qubit gates" defined as

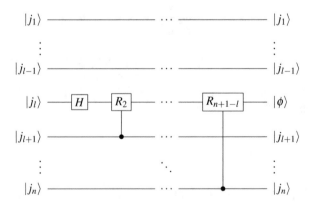

Fig. 4.6 Part of the quantum Fourier transform circuit that acts on qubit $|j_l\rangle$. The value of all qubits does not change, except $|j_l\rangle$ that changes to $|\phi\rangle = \frac{1}{\sqrt{2}}|0\rangle + \exp\left(\frac{2\pi i j}{2^{n+1-l}}\right)|1\rangle$

$$CR_k = \begin{bmatrix} 1 & 0 \\ 0 & \exp\left(2\pi i \, \frac{j_{k+l-1}}{2^k}\right) \end{bmatrix}. \tag{4.38}$$

In fact, CR_k is a two-qubit gate that is temporarily being represented as an one-qubit gate.

In order to simplify the calculations, notice that

$$H |j_l\rangle = \frac{|0\rangle + e^{2\pi i \frac{j_l}{2}} |1\rangle}{\sqrt{2}}$$

$$= CR_1 |+\rangle, \tag{4.39}$$

where $|+\rangle = \frac{1}{\sqrt{2}}(|0\rangle + |1\rangle)$. Thus, instead of using the equation

$$|\psi\rangle = CR_{n+1-l} \dots CR_2 H |j_l\rangle, \tag{4.40}$$

which can be read directly from Fig. 4.6, we will use

$$|\psi\rangle = CR_{n+1-l} \dots CR_2 CR_1 |+\rangle. \tag{4.41}$$

We define the operator

$$PR_{n+1-l} = \prod_{k=n+1-l}^{1} CR_k, \tag{4.42}$$

where the product is in reversed order—remember this is a product of unitary operators, and thus non-commutative. From Eqs. (4.38) and (4.42), we get

$$PR_{n+1-l} = \frac{1}{\sqrt{2}} \begin{bmatrix} 1 & 0 \\ 0 & \exp 2\pi i \left(\frac{j_n}{2^{n+1-l}} + \dots + \frac{j_l}{2}\right) \end{bmatrix}$$

$$= \frac{1}{\sqrt{2}} \begin{bmatrix} 1 & 0 \\ 0 & \exp\left(2\pi i \, \frac{j}{2^{n+1-l}}\right) \end{bmatrix}, \tag{4.43}$$

where we have used that $j = \sum_{m=1}^{n} j_m 2^{n-m}$ and the fact that the first $l-1$ terms of this expansion do not contribute—they are integer multiples of $2\pi i$ in Eq. (4.43). We finally get

$$|\psi\rangle = PR_{n+1-l} |+\rangle$$

$$= \frac{1}{\sqrt{2}} \left(|0\rangle + e^{2\pi i \frac{j}{2^{n+1-l}}} |1\rangle\right). \tag{4.44}$$

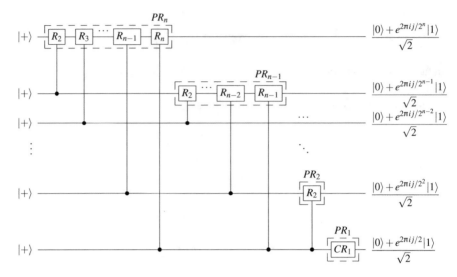

Fig. 4.7 Intermediate circuit for the quantum Fourier Transform. The input is taken as $|+\rangle$ for calculation purposes as explained in Eq. (4.39). The output is in reverse order with respect to Eq. (4.35)

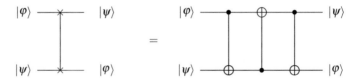

Fig. 4.8 Decomposition of swap gate in terms of universal gates

Notice that PR_{n+1-l} cannot be implemented directly acting only in the l-th qubit, because it requires the values of j_{l+1} to j_n.

The next step is the circuit depicted in Fig. 4.7. We have merged the R_k gates using Eq. (4.42). The gates PR_k, with k from n to 1, are placed in sequence in Fig. 4.7, so that the output of the first qubit is the last term of Eq. (4.35), corresponding to the action of PR_n on $|\psi_1\rangle$ controlled by the other qubits, which do not change. The same process is repeated by PR_{n-1} acting on $|\psi_2\rangle$, yielding the term before the last in Eq. (4.35), and so on, until reproducing all the terms of the Fourier transform. Now it remains to reverse the order of the states of the qubits.

In order to reverse the states of two generic qubits, we use the circuit of Fig. 4.8. Let us show why this circuit works as desired. Take as input a vector of the computation basis, say, $|\varphi\rangle |\psi\rangle = |0\rangle |1\rangle$. The first CNOT gate of Fig. 4.8 does not change this state. Then, the second CNOT gate—which is upside down—changes it to $|1\rangle |1\rangle$. Finally, the last CNOT gate changes the state to $|1\rangle |0\rangle$. The output is $|\psi\rangle |\varphi\rangle$, which means that the circuit worked for basis state $|0\rangle |1\rangle$ as input. If we repeat the same process with $|0\rangle |0\rangle$, $|1\rangle |0\rangle$, and $|1\rangle |1\rangle$ as inputs, we conclude that

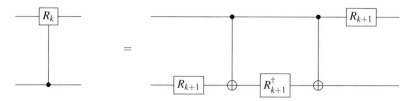

Fig. 4.9 Decomposition of the controlled R_k gates in terms of the universal gates

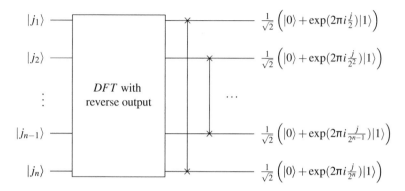

Fig. 4.10 The complete circuit for the quantum Fourier Transform

the circuit inverts all states of the computational basis, therefore it inverts a generic state of the form $|\varphi\rangle\,|\psi\rangle$.

The decomposition is still not complete. It remains to write the controlled R_k gates in terms of CNOT and one-qubit gates. This decomposition is given in Fig. 4.9. The verification of this decomposition is straightforward. One simply follows what happens to each vector of the computational basis $\{|00\rangle, |01\rangle, |10\rangle, |11\rangle\}$ in both circuits.

The complete circuit for the quantum Fourier transform is given in Fig. 4.10. Now we can calculate the complexity of the quantum Fourier circuit. By counting the number of elementary gates in Figs. 4.6, 4.7, 4.8 and 4.9, we get the leading term $5n^2/2$, which implies that the complexity is $O(n^2)$.

By now one should be asking about the decomposition of V_x in terms of the elementary gates. Operator V_x is the largest gate of Fig. 4.1 and the "bottleneck" of the quantum factoring algorithm, due to the time and space consumed to perform the modular exponentiation. However, this bottleneck is not so strict, since by using the well known classical method of repeated squaring and ordinary multiplication algorithms, the complexity to calculate modular exponentiation is $O(n^3)$. The quantum circuit can be obtained from the classical circuit by replacing the irreversible classical gates by the reversible quantum counterpart.

For simplicity, when we discussed operator V_x we kept x fixed. This approach is a problem in recursive calls of the algorithm when x changes, since for each x, a new circuit must be built. In order to be practical, the circuit for operator V_x should be designed in such a way that the calculations are performed for an arbitrary x, which in turn should be provided through an ancillary register.

4.6   Computational Simulations

The *LIQUiD* programming language[8] has been developed by Microsoft, and allows advanced simulations of several quantum algorithms, including a builtin test of Shor's algorithm. The user just needs to tell the number to be factorized and whether the circuit should be optimized. For instance, if one wants to factorize $N = 21$ with the circuit just studied in this chapter, without optimization, the command to be executed should be

```
1  $ mono Liquid.exe "__Shor(21, false )"
```

on Linux or MacOS operating systems, or

```
1  > Liquid.exe __Shor(21, false )
```

on Microsoft Windows. In either case, the output would inform several details concerning time and memory consumption besides the result.

Sympy is a Python library for symbolic mathematics which has already some commands for simulating Shor's algorithm.[9] The simulation can be performed as easily as

```
1  import sympy.physics.quantum.shor as q
2  q.shor(21)
```

The result is a factor of N if the random number leads to a successful run, and an error message otherwise. The output does not give much details, although the source code is open and much more interesting as a learning resource.

The Quantum Computing Playground[10] contains a predefined open-source simulation of Shor's algorithm, and provides several visualizations. The source code for this simulation uses a language called QScript, and the website gives a tutorial teaching how to use it.

There is also an implementation of Shor's algorithm for IBM Quantum Experience in [3].

References

1. Boneh, D., Lipton, R.J.: Quantum cryptanalysis of hidden linear functions. In: Annual International Cryptology Conference, pp. 424–437. Springer, Heidelberg (1995)
2. Childs, A.M., van Dam, W.: Quantum algorithms for algebraic problems. Rev. Mod. Phys. **82**, 1–52 (2010). https://doi.org/10.1103/RevModPhys.82.1

---

[8] Available at http://stationq.github.io/Liquid/ as of October 2016.

[9] Available at http://docs.sympy.org/dev/modules/physics/quantum/shor.html as of October 2016.

[10] Available at http://www.quantumplayground.net as of October 2016.

3. Coles, P.J., Eidenbenz, S., Pakin, S., Adedoyin, A., Ambrosiano, J., Anisimov, P., Casper, W., Chennupati, G., Coffrin, C., Djidjev, H., Gunter, D., Karra, S., Lemons, N., Lin, S., Lokhov, A., Malyzhenkov, A., Mascarenas, D., Mniszewski, S., Nadiga, B., O'Malley, D., Oyen, D., Prasad, L., Roberts, R., Romero, P., Santhi, N., Sinitsyn, N., Swart, P., Vuffray, M., Wendelberger, J., Yoon, B., Zamora, R., Zhu, W.: Quantum Algorithm Implementations for Beginners. Technical Report LA-UR:2018-229, Los Alamos National Laboratory, Los Alamos (2018)

4. Cooley, J.W., Tukey, J.W.: An algorithm for the machine calculation of complex Fourier series. Math. Comput. **19**(90), 297 (1965). https://doi.org/10.2307/2003354

5. Crandall, R., Pomerance, C.: Prime Numbers: A Computational Perspective, 2nd edn. Springer, New York (2005). https://doi.org/10.1007/0-387-28979-8

6. Einarsson, G.: Probability Analysis of a Quantum Computer (2003). http://arxiv.org/abs/quant-ph/0303074

7. Ekert, A., Jozsa, R.: Quantum computation and Shor's factoring algorithm. Rev. Mod. Phys. **68**(3), 733–753 (1996). https://doi.org/10.1103/RevModPhys.68.733

8. Gathen, J.V.Z., Gerhard, J.: Modern Computer Algebra. Cambridge University Press, Cambridge (1999)

9. Hoffstein, J., Pipher, J., Silverman, J.: An Introduction to Mathematical Cryptography. Undergraduate Texts in Mathematics. Springer, New York (2008). https://doi.org/10.1007/978-0-387-77993-5

10. Jozsa, R.: Quantum algorithms and the Fourier transform. Proc. R. Soc. Lond. A Math. Phys. Eng. Sci. **454**(1969), 323–337 (1998). https://doi.org/10.1098/rspa.1998.0163

11. Jozsa, R.: Quantum factoring, discrete logarithms, and the hidden subgroup problem. Comput. Sci. Eng. **3**(2), 34–43 (2001). https://doi.org/10.1109/5992.909000

12. Kitaev, A.Y.: Quantum measurements and the abelian stabilizer problem (1995). ArXiv: quant-ph/9511026

13. Knuth, D.E.: The Art of Computer Programming, vol. 2, Seminumerical Algorithms, 3rd edn. Addison-Wesley, Upper Saddle River (1997)

14. Lenstra, H.W.: Factoring integers with elliptic curves. Ann. Math. **126**(3), 649–673 (1987)

15. Lomonaco, S.J.: Quantum computation: a grand mathematical challenge for the twenty-first century and millennium. In: Samuel, J., Lomonaco, J. (ed.) Proceedings of Symposia in Applied Mathematics, pp. 161–179. American Mathematical Society, Washington (2000). https://doi.org/10.1090/psapm/058/1922897

16. Marquezino, F., Portugal, R., Sasse, F.: Obtaining the quantum Fourier transform from the classical FFT with QR decomposition. J. Comput. Appl. Math. **235**(1), 74–81 (2010). https://doi.org/10.1016/j.cam.2010.05.012

17. Montanaro, A.: Quantum algorithms: an overview. Npj Quantum Inf. **2**(15023) (2016). https://doi.org/10.1038/npjqi.2015.23

18. Monz, T., Nigg, D., Martinez, E.A., Brandl, M.F., Schindler, P., Rines, R., Wang, S.X., Chuang, I.L., Blatt, R.: Realization of a scalable shor algorithm. Science **351**(6277), 1068–1070 (2016). https://doi.org/10.1126/science.aad9480

19. Nielsen, M.A., Chuang, I.L.: Quantum Computation and Quantum Information. Cambridge University Press, Cambridge (2010). https://doi.org/10.1017/CBO9780511976667

20. Pollard, J.M.: A monte carlo method for factorization. BIT Numer. Math. **15**(3), 331–334 (1975). https://doi.org/10.1007/BF01933667

21. Shor, P.: Algorithms for quantum computation: discrete logarithms and factoring. In: Proceedings 35th Annual Symposium on Foundations of Computer Science, pp. 124–134. IEEE Computer Society, Washington (1994). https://doi.org/10.1109/SFCS.1994.365700

22. Shor, P.W.: Polynomial-time algorithms for prime factorization and discrete logarithms on a quantum computer. SIAM Rev. **41**(2), 303–332 (1999). https://doi.org/10.1137/S0036144598347011

23. Vandersypen, L.M.K., Steffen, M., Breyta, G., Yannoni, C.S., Sherwood, M.H., Chuang, I.L.: Experimental realization of Shor's quantum factoring algorithm using nuclear magnetic resonance. Nature **414**(6866), 883–887 (2001). https://doi.org/10.1038/414883a

24. Wagstaff Jr., S.S.: The joy of factoring. In: Student Mathematical Library, vol. 68. American Mathematical Society, Washington (2013). https://doi.org/10.1090/stml/068

Chapter 5
Quantum Walks

In order to design new algorithms, we usually consider patterns or techniques that have already been successfully applied to similar problems. Every student of computer science must learn a large repertoire of such techniques. In the previous chapters, we learned the first patterns for quantum algorithm design, namely the amplitude amplification and the quantum Fourier transform. In this chapter, we study another technique, analogous to the classical random walk: the quantum walk.

The field of quantum walks is too large to be fully explained in just one chapter. Therefore, we focus on two models of quantum walks: the coined model and the staggered model. The coined model was the first quantum walk model to be defined, still in 1993. The staggered model is much more recent and generalizes the most important types of quantum walks.

5.1 Classical Random Walks

Random walks have a myriad of applications in several areas of knowledge, such as chemistry, biology, economics, psychology and so on. In particular, they are a successful technique used by computer scientists in the design of algorithms. A classical example of algorithm based on a random walk is Schöning's algorithm for the k-SAT problem—that is, the Boolean k-Satisfiability Problem. This algorithm initially guesses an assignment for the Boolean expression and then tries to find a satisfying assignment by successively flipping a random literal from a random unsatisfied clause.

We may define a random walk as the mathematical description of a particle (or walker) that moves across discrete positions in the space through a succession of random steps. In the above example of Schöning's algorithm, the "particle" assumes positions corresponding to each possible assignment for the Boolean expression. Flipping the value of a literal, in this example, is equivalent to moving the particle to a neighbor position.

© The Author(s), under exclusive license to Springer Nature Switzerland AG 2019
F. de Lima Marquezino et al., *A Primer on Quantum Computing*, SpringerBriefs in
Computer Science, https://doi.org/10.1007/978-3-030-19066-8_5

Table 5.1 Probability of finding the walker at position n after t steps of the classical random walk, assuming an unbiased coin, and particle initially located at origin. Empty cells represent zero probability

	Position								
Step	-4	-3	-2	-1	0	$+1$	$+2$	$+3$	$+4$
0					1				
1				1/2		1/2			
2			1/4		1/2		1/4		
3		1/8		3/8		3/8		1/8	
4	1/16		1/4		3/8		1/4		1/16

The idea of a random walk becomes much clearer when we take a toy example. Let us consider a classical particle that moves on discrete positions of the line, the direction of motion being determined by the toss of a coin. The position of the particle at any time is given by $n \in \mathbb{Z}$. If the coin toss gives heads the particle moves to $n + 1$, and if it gives tails the particle moves to $n - 1$. Assuming a non-biased coin and that the particle is initially located at the origin, $n = 0$, then at time $t = 1$ we can find the particle at position $n = -1$ with probability $1/2$, or $n = 1$ with probability $1/2$. And if we toss the coin and move the particle once again, then at time $t = 2$ we can find the particle at position $n = -2$ with probability $1/4$, or $n = 0$ with probability $1/2$, or at position $n = 2$ with probability $1/4$. If we repeat this process over and over, we will observe a probability distribution such as the one in Table 5.1.

Let us denote by $p_t(n)$ the probability of the particle being at a given position n at time t. One can prove that

$$p_t(n) = \begin{cases} \frac{1}{2^t} \binom{t}{\frac{t+n}{2}} & \text{if } t + n \text{ is even and } n \leq t \\ 0 & \text{otherwise,} \end{cases} \tag{5.1}$$

where $\binom{t}{\frac{t+n}{2}} = \frac{a!}{(a-b)!b!}$. For a fixed t, the previous expression is a *binomial distribution*. If we plot $p_t(n)$ for a sufficiently large fixed t and omit the points with probability zero, we obtain a bell-shaped curve. This curve is symmetric across the $n = 0$ axis, its height decreases and its width increases as a function of time t.

Several questions can be asked about the process that we just described. For instance, we may want to determine how far from the origin the particle can be found at a time t, and how this *expected distance* changes as a function of time. This statistical quantity is captured by the *standard deviation* of the position. Since $p_t(n) = p_t(-n)$, the *average position* is

$$\langle n \rangle = \sum_{n=-\infty}^{\infty} n p_t(n) = 0, \tag{5.2}$$

and the standard deviation of the position is

$$\sqrt{\langle n^2 \rangle - \langle n \rangle^2} = \sqrt{t}. \tag{5.3}$$

The idea presented in this section can be easily extended to describe random walks on more general graphs, with an elegant mathematical description. For a complete description of the theory of classical random walks the reader may refer to Spitzer [10], or to Lawler and Limic [4].

5.2 Coined Quantum Walks

Now we want to formulate a quantum version for the random walk that we described in the previous section. Clearly, the mathematical description of the quantum walk must obey the postulates of quantum mechanics. First of all, we need to know which Hilbert space will represent the state vector. Since we are studying the toy example of a particle over the line, it is tempting to propose an infinite-dimensional Hilbert space with basis vector from $\{|n\rangle, n \in \mathbb{Z}\}$, so that we can represent the position of the quantum walker. However, this solution would not be good enough because the only allowed unitary evolutions would be trivial—only right or left shifts would be possible, up to a phase [5].

The first effective model of quantum walk that we will present is the *coined quantum walk*, which was first proposed by Aharonov, Davidovich and Zagury [2]. In this model, we need two Hilbert subspaces: one to represent the position of the walker, and one to represent an additional degree of freedom which we may interpret as a *coin* state. We may replace the coin from the previous description of a random walk by a two-dimensional quantum system, associating heads with state $|0\rangle$ and tails with state $|1\rangle$, for instance. We may also associate the position of the particle with a state $|n\rangle$, $n \in \mathbb{Z}$, for instance. Therefore, the state of the walker at any given time is given by a unit vector $|\Psi(t)\rangle \in \mathcal{H}_C \otimes \mathcal{H}_P$, where $\mathcal{H}_C$ is the two-dimensional coin subspace, and $\mathcal{H}_C$ is the infinite-dimensional position subspace.

In this new set-up we need two unitary operators: first, a coin operator C that changes the coin state mimicking the coin toss from our previous description, and then a shift operator S that changes the position state depending on the coin state. The coin operator can be any one that matches the dimension of the coin subspace. However, a Hadamard gate is very convenient for the quantum walk on the line, so we will adopt it in this section. In order to keep the consistency with our previous descriptions, let us define the shift operator as

$$S|j\rangle|x\rangle = |j\rangle\Big|x + (-1)^j\Big\rangle, \tag{5.4}$$

for all $j \in \{0, 1\}$, and for all $x \in \mathbb{Z}$. Notice that operator S shifts the walker to the right if the coin is $j = 0$, and to the left if the coin is $j = 1$. Equation (5.4) is just an example, and other unitary shift operators are also possible.

The quantum operator that describes one step of the quantum walk is

$$U = S(C \otimes I), \tag{5.5}$$

where the identity operator I is over the position subspace. The quantum walker evolves according to successive applications of operator U, without intermediate measurements[1]—that is, the state of the quantum walker at a given time t is given by

$$|\Psi(t)\rangle = U^t|0\rangle. \tag{5.6}$$

Let us consider a scenario where the coin is initially set to state $|0\rangle$ and the walker is initially located at position $|0\rangle$—that is, the state of the walker is given by $|\Psi(0)\rangle = |0\rangle|0\rangle$. Then, applying the evolution operator U we obtain

$$|\Psi(1)\rangle = \frac{|0\rangle|+1\rangle + |1\rangle|-1\rangle}{\sqrt{2}}. \tag{5.7}$$

If we measure the position, we find the walker at position -1 or $+1$ with equal probability, which is the same distribution observed in the classical random walk. However, if instead of measuring the position, we apply the evolution operator again, then this time we get

$$|\Psi(2)\rangle = \frac{|0\rangle|+2\rangle + |0\rangle|0\rangle + |1\rangle|0\rangle - |1\rangle|-2\rangle}{2}. \tag{5.8}$$

If we measure the position, we find the walker at position -2 with probability $1/4$, or in position 0 with probability $1/2$, or at position $+2$ with probability $1/4$. That is still the same probability distribution as in the classical random walk.

It seems that the quantum walk is not working as we expected. However, notice that the state in Eq. (5.8) has a minus sign! If we insist and apply the evolution operator one more time, we finally get

$$|\Psi(3)\rangle = \frac{|0\rangle|+3\rangle + 2|0\rangle|1\rangle + |1\rangle|1\rangle - |0\rangle|-1\rangle + |1\rangle|-3\rangle}{2\sqrt{2}}. \tag{5.9}$$

If we measure the position now, we find the walker at position -3 with probability $1/8$, or in position -1 with probability $1/8$, or in position $+1$ with probability $5/8$, or in position $+3$ with probability $1/8$. This is different from the classical

---

[1]If we measure the quantum walker after each step, then we recover the probability distribution of the classical random walker.

Table 5.2 Probability of finding the walker at position n after t steps of the quantum walk, assuming the Hadamard coin operator, and particle initially located at origin. Empty cells represent zero probability

Step	Position								
	−4	−3	−2	−1	0	+1	+2	+3	+4
0					1				
1				1/2		1/2			
2			1/4		1/2		1/4		
3		1/8		1/8		5/8		1/8	
4	1/16		1/8		1/8		5/8		1/16

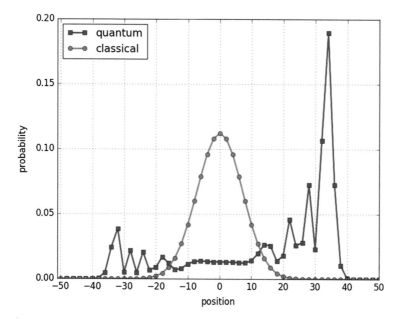

Fig. 5.1 Probability distributions of the positions of a classical random walk, and a quantum walk, after $t = 50$ steps, assuming the particle initially located at origin. We omit sites with zero probability for clarity

random walk, and asymmetric! If we repeat this process over and over, without intermediate measurements, we will observe a probability distribution such as the ones in Table 5.2 or in Fig. 5.1. One can also obtain a symmetrical distribution in the quantum walk—see Exercise 5.3.

Exercise 5.1 Verify the probability distributions at steps $t = 3$ and $t = 4$ from Table 5.2.

Exercise 5.2 Find again the entries of Table 5.2, however assuming

$$|\Psi(0)\rangle = |1\rangle|0\rangle \tag{5.10}$$

as the initial condition.

Exercise 5.3 Find again the entries of Table 5.2, however assuming

$$|\Psi(0)\rangle = \left(\frac{|0\rangle + i|1\rangle}{\sqrt{2}}\right)|0\rangle \tag{5.11}$$

as the initial condition.

5.2.1 Generalizing by Means of Examples

The ideas presented in the previous toy example can be easily extended to any other undirected simple graph—and even to some instances of directed graphs, even though the description of this case would go beyond the scope of this book. The extension to regular graphs is easier to grasp, so we will start from there. Consider a d-regular graph $\Gamma(V, E)$, where V is the vertex set and E is the edge set. A quantum walker on Γ can be modeled by a Hilbert space $\mathscr{H}_C \otimes \mathscr{H}_P$, where the coin subspace $\mathscr{H}_C$ has dimension corresponding to the degree d of the graph, and the position subspace $\mathscr{H}_P$ has dimension corresponding to the order $|V|$ of the graph.

Following the ideas presented previously, we may now want to define a coined quantum walk on the infinite two-dimensional lattice. In this case, we should consider a coin subspace $\mathscr{H}_C$ spanned by the basis $\{|0, 0\rangle, |0, 1\rangle, |1, 0\rangle, |1, 1\rangle\}$, and a position subspace $\mathscr{H}_P$ spanned by the basis $\{|x, y\rangle, -\infty \leq x, y \leq +\infty\}$. If instead of the infinite lattice we wanted to model a quantum walk on the torus, we would replace the position basis by $\{|x, y\rangle, -N \leq x, y \leq +N\}$, where N is a positive integer. If we wanted to model a coined quantum walk on the d-dimensional hypercube, we should consider a coin subspace $\mathscr{H}_C$ spanned by the basis $\{|0\rangle, |1\rangle, \ldots, |d-1\rangle\}$, and a position subspace $\mathscr{H}_P$ spanned by the basis $\{|0\rangle, |1\rangle, \ldots, |2^d - 1\rangle\}$. As we shall see later on, it may be convenient to label the position vectors of the hypercube in binary notation, i.e., $|x\rangle \equiv |x_{d-1}x_{d-2}\cdots x_0\rangle$, where $0 \leq x_j \leq 1$ for every j.

Defining the coin operator is straightforward, since any unitary operator over $\mathscr{H}_C$ is allowed. In our first toy example—the coined quantum walk on the line—we chose the Hadamard matrix $H = \frac{1}{\sqrt{2}}\begin{bmatrix} 1 & 1 \\ 1 & -1 \end{bmatrix}$ as coin. For coined quantum walks on two-dimensional lattices, we may choose $H \otimes H$ as coin operator—or any other 4×4 unitary matrix. On the other hand, the Hadamard operator would not work in general for quantum walks on d-dimensional hypercubes. However, one can always take as a convenient coin operator the Grover operator

$$G = 2\left|s^C\right\rangle\left\langle s^C\right| - I, \tag{5.12}$$

where $\left|s^C\right\rangle = \frac{1}{\sqrt{d}}\sum_{j=0}^{d-1}|j\rangle$. The Grover coin operator can be defined for coin subspaces of any dimension. Moreover, the Grover coin is very important for search algorithms, where it is usually the most efficient choice.

The shift operator must not only be unitary, but also respect the neighborhood of the vertices. In the quantum walk on the infinite two-dimensional lattice, we may choose the shift operator

$$S|j, k\rangle|x, y\rangle = |j, k\rangle \Big| x + (1 - j)(-1)^k, y + j(-1)^k \Big\rangle, \qquad (5.13)$$

where $|j, k\rangle$ represents the coin state, and $|x, y\rangle$ represents the position state.[2] Notice that j, in this example, indicates whether the walker will be shifted across the vertical or the horizontal axis, and k indicates whether the walker is shifted towards the positive or negative direction. In this case, the coin subspace is left untouched after the shift operation. Another valid choice of shift operator would be

$$S|j, k\rangle|x, y\rangle = |j, 1 - k\rangle \Big| x + (1 - j)(-1)^k, y + j(-1)^k \Big\rangle, \qquad (5.14)$$

which flips the direction after each shift operation. This kind of shift operation is know as flip-flop, and is very important for search algorithms. In the quantum walk on the d-dimensional hypercube, we may choose a shift operator such as

$$S|d\rangle|x_{d-1}, x_{d-2}, \cdots, x_d, \cdots, x_0\rangle = |d\rangle|x_{d-1}, x_{d-2}, \cdots, 1 - x_d, \cdots, x_0\rangle, \qquad (5.15)$$

which flips just one bit of the position state corresponding to the coin state. Since two consecutive bit flips cancel each other out, the shift operator from Eq. (5.15) is already a flip-flop shift.

In general, we can define a valid shift operator by the following procedure. For each vertex $v \in V$, let us label its incident edges with $j \in \{0, 1, \ldots, d - 1\}$. For each labeled edge j incident do v, if the other endpoint is w, then the shift operator should act as

$$S|j\rangle|v\rangle = |j\rangle|w\rangle. \qquad (5.16)$$

For consistency, we would still like to have

$$S|j\rangle|w\rangle = |j\rangle|v\rangle, \qquad (5.17)$$

meaning that we go from vertex v to vertex w with the same edge used to go from vertex w to vertex v.

Sometimes, it is not possible to simultaneously satisfy Eqs. (5.16) and (5.17). For example, we know from Vizing's theorem [11] that there are examples of graphs that need $\Delta + 1$ different values to label its edges, where Δ is the maximum degree of the graph.

---

[2]For the two-dimensional torus, we easily can adapt Eq. (5.13) by using modular arithmetics.

If the graph is irregular, then we can conditionally apply a different coin operator, with the adequate dimension, to each of the vertices. In this case, we can no longer represent the coin operator as a separable operation, as in Eq. (5.5). We should instead write the evolution operator as

$$U = SC'. \tag{5.18}$$

5.3   Staggered Quantum Walks

A quantum walk on a graph is called a *staggered quantum walk* when it is based on a *tessellation cover*. A *tessellation cover* is a set of tessellations that covers the edges of the graph and a *tessellation* is a partition of the graph into *cliques*.[3] The staggered model was initially devised to generalize coinless quantum walks on lattices [8], which were obtained by transforming the coin internal space into extra spatial degrees of freedom. This is accomplished by expanding the graph by adding new vertices so that the possible coin values are represented by vertices in the expanded graph. This idea can be extended to a general graph with the use of a tessellation cover of size two in such way to include the entire flip-flop coined model. On the other hand, due to the extensiveness of the construction, the staggered model has also included all Szegedy's quantum walks as particular cases. All those particular models are obtained using two tessellations.[4] Staggered quantum walks with more than two tessellations provide more versatility to the model, admit perfect state transfer, and can be used to discretize the continuous-time model in some graph instances [6].

5.3.1   Tessellation Cover and the Evolution Operator

Consider a graph $\Gamma(V, E)$, where V is the vertex set and E is the edge set. To simplify the discussion, we suppose that the graph is finite. A tessellation $\mathscr{T}$ of a graph $\Gamma(V, E)$ is a partition of V into cliques, that is, $\mathscr{T}$ is a set of cliques P_i, called polygons, such that the polygons are pairwise non-intersecting and the union $\cup_{i=1}^{p} P_i$ is the vertex set, where p is the tessellation size (number of polygons). An edge e belongs to $\mathscr{T}$ if the vertices incident to e belong to the same polygon.

A tessellation cover $\mathscr{C} = \{\mathscr{T}_1, \ldots, \mathscr{T}_k\}$ of size k is a set of tessellations such that the union $\cup_{j=1}^{k} \mathscr{E}(\mathscr{T}_j)$ is the edge set, where $\mathscr{E}(\mathscr{T}_j)$ is the set of edges belonging

---

[3]A clique a graph $\Gamma(V, E)$ is a subset of V such that any two vertices in the subset are adjacent, that is, a clique induces a complete subgraph.

[4]It is possible to strengthen the power of the coined model by using the square of the evolution operator [3].

Fig. 5.2 A diamond ring covered by two tessellations

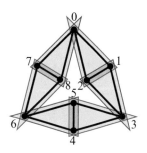

to $\mathscr{T}_j$. A graph is called k-tessellable if there exists a tessellation cover of size k. Given a graph, an interesting problem is to find the minimum size of a tessellation cover [1].

Consider for instance the tessellation cover of the graph depicted in Fig. 5.2. The graph is a *diamond ring*, where a diamond is defined as a 4-clique minus one edge. Tessellation $\mathscr{T}_1$ (red) is $\{\{0, 1, 2\}, \{3, 4, 5\}, \{6, 7, 8\}\}$ and tessellation $\mathscr{T}_2$ (blue) is $\{\{0, 7, 8\}, \{1, 2, 3\}, \{4, 5, 6\}\}$. Each polygon has size 3 and each tessellation has 3 polygons (size 3). The edges in each tessellation are $\mathscr{E}(\mathscr{T}_1) = \{(0, 1), (0, 2), (1, 2), (3, 4), (3, 5), (4, 5), (6, 7), (6, 8), (7, 8)\}$ in the red tessellation and $\mathscr{E}(\mathscr{T}_2) = \{(0, 7), (0, 8), (1, 2), (1, 3), (2, 3), (4, 5), (4, 6), (5, 6), (7, 8)\}$ in the blue tessellation. It is straightforward to check that the tessellation cover contains all edges, that is, $\mathscr{E}(\mathscr{T}_1) \cup \mathscr{E}(\mathscr{T}_2) = E$.

Let $\mathscr{H}^{|V|}$ be a $|V|$-dimensional Hilbert space whose computational basis is $\{|v\rangle : v \in V\}$, that is, the computational basis is indexed by the vertices. Consider a tessellation cover $\mathscr{C} = \{\mathscr{T}_1, \ldots, \mathscr{T}_k\}$ of size k. The evolution operator $\mathscr{U}$ of the staggered model is the product of k unitary operators

$$\mathscr{U} = U_k \ldots U_2 U_1, \tag{5.19}$$

where

$$U_j = \mathrm{e}^{i\theta H_j}, \tag{5.20}$$

θ is an angle and H_j is a local Hermitian operator associated with tessellation $\mathscr{T}_j$. The tessellation ordering usually matters because the unitary operators are usually pairwise non-commuting. Given a tessellation $\mathscr{T}_j$, there are different ways to define operator H_j. We will start by describing the most general association, which can be used to reproduce any instance of the flip-flop coined quantum walk, and we proceed to a particular case, which reproduces any instance of the Szegedy model [8].

We can associate a graph tessellation $\mathscr{T}$ of size p with a Hermitian operator H in $\mathscr{H}^{|V|}$ by demanding the following restriction: H is a sum of p Hermitian matrices so that each matrix is associated with a polygon P in the sense that the matrix acts non-trivially only on the vector components that correspond to the vertices in P. By construction, H is a local operator, that is, if the walker is on a vertex v of a polygon

P, the action of H on $|v\rangle$ outputs a vector that is a superposition of vertices in P. Since P is a clique, the action of H moves the walker inside a neighborhood of v. Notice that $U = \exp(i\theta H)$ is also local.

An important subcase is obtained when we impose the following extra restriction: $H^2 = I$. Under this extra restriction we have

$$U = \cos(\theta)\, I + i\, \sin(\theta)\, H. \tag{5.21}$$

There is an interesting method to obtain this subcase. We associate each polygon P^ℓ with vector

$$\left|P^\ell\right\rangle = \frac{1}{\sqrt{\left|P^\ell\right|}} \sum_{v \in P^\ell} |v\rangle, \tag{5.22}$$

where $\left|P^\ell\right|$ is the number of vertices in P^ℓ. Now we define H in the following way:

$$H = 2 \sum_{\ell=1}^{p} \left|P^\ell\right\rangle\!\left\langle P^\ell\right| - I. \tag{5.23}$$

By construction, H is Hermitian and unitary. Therefore, $H^2 = I$. Notice that this process must be repeated for each tessellation $\mathscr{T}_j$ of the tessellation cover $\mathscr{C} = \{\mathscr{T}_1, \ldots, \mathscr{T}_k\}$. Any instance of Szegedy's model on a bipartite graph G is reproduced in this context up to a global phase by a staggered quantum walk on the *line graph*[5] of G and by taking $\theta = \pi/2$ and $k = 2$ [9].

Exercise 5.4 Find the maximal cliques of a N-circle and show that this graph is 2-tessellable if N is even and is 3-tessellable if N is odd. Find the *clique graph*[6] a N-circle and show the clique graph is 2-colorable[7] if N is even and is 3-colorable if N is odd.

Exercise 5.5 Find a minimum tessellation cover for the Hajós graph (a triangle surrounded by three triangles each one sharing an edge in common with the central triangle) and show that at least one maximal clique is contained in no tessellation of any minimum tessellation cover. Show the clique graph of the Hajós graph is 4-colorable.

---

[5] A line graph of a graph Γ (called root graph) is another graph $L(\Gamma)$ so that each vertex of $L(\Gamma)$ represents an edge of Γ and two vertices of $L(\Gamma)$ are adjacent if and only if their corresponding edges share a common vertex in Γ.

[6] A clique graph $K(G)$ of a graph G is a graph such that every vertex represents a maximal clique of G and two vertices of $K(G)$ are adjacent if and only if the corresponding maximal cliques in G share at least one vertex in common.

[7] A k-colorable graph is the one whose vertices can be colored with at most k colors so that no two adjacent vertices share the same color.

Exercise 5.6 The *wheel graph* is the graph W_n for $n > 2$ with vertex set $\{0, 1, 2, \ldots, n\}$ and edge set $\{\{0, n\}, \{1, n\}, \ldots, \{n - 1, n\}, \{0, 1\}, \{1, 2\}, \ldots, \{n - 2, n - 1\}, \{n - 1, 0\}\}$. Show that W_n is $(n/2)$-tessellable if n is even. Show that the chromatic number of the clique graph $K(W_n)$ is n. By adding new edges to the wheel graph, try to provide examples of graph classes that are $(n/3)$-tessellable such that the chromatic number of the clique graph of graphs in this new class is still n. Can you provide an example of a 3-tessellable graph class with chromatic number of the clique graph equal to n?

Exercise 5.7 Show that if vectors $\left| P^\ell \right\rangle$ are associated with polygons P^ℓ of a tessellation $\mathscr{T}$ then $\langle P^\ell | P^{\ell'} \rangle = \delta_{\ell \ell'}$. Show that operator H defined in Eq. (5.23) is Hermitian and unitary. Show that $H^2 = I$.

Exercise 5.8 Consider the complete graph G with N vertices. Let $\mathscr{T}$ a tessellation of G consisting of a single set that cover all vertices. Prove that operator H defined in Eq. (5.23) is the Grover operator.

Exercise 5.9 Prove that if $H^2 = I$ then $\exp(i\theta H) = \cos(\theta)\, I + i \sin(\theta)\, H$ for any matrix H.

5.3.2 An Example Using the Diamond Ring of Size 9

In this subsection, we analyze in details the staggered quantum walk on the diamond ring of size $N = 9$ ($n = 3$ diamonds) depicted in Fig. 5.2 and then we generalize for a diamond ring of any size $N = 3n$ in the next subsection.

Following the recipe given in Sect. 5.3.1, Hamiltonian H_1, which is associated with tessellation $\mathscr{T}_1$, must be the sum of three Hermitian matrices $H_1^{(1)}, H_1^{(2)}, H_1^{(3)}$ because tessellation $\mathscr{T}_1$ has three polygons. Let us define the first one as

$$H_1^{(1)} = \frac{1}{3} \begin{bmatrix} -1 & 2 & 2 & 0 & 0 & 0 & 0 & 0 & 0 \\ 2 & -1 & 2 & 0 & 0 & 0 & 0 & 0 & 0 \\ 2 & 2 & -1 & 0 & 0 & 0 & 0 & 0 & 0 \\ 0 & 0 & 0 & 0 & 0 & 0 & 0 & 0 & 0 \\ 0 & 0 & 0 & 0 & 0 & 0 & 0 & 0 & 0 \\ 0 & 0 & 0 & 0 & 0 & 0 & 0 & 0 & 0 \\ 0 & 0 & 0 & 0 & 0 & 0 & 0 & 0 & 0 \\ 0 & 0 & 0 & 0 & 0 & 0 & 0 & 0 & 0 \\ 0 & 0 & 0 & 0 & 0 & 0 & 0 & 0 & 0 \end{bmatrix}. \tag{5.24}$$

Notice that $H_1^{(1)}$ is Hermitian and acts non-trivially only on the subspace spanned by the vertices that are in polygon $P_1^{(1)} = \{0, 1, 2\}$. Any other choice that yields a Hermitian matrix that has nonzero elements only in the entries corresponding the vertices of $P_1^{(1)}$ is acceptable. Matrices $H_1^{(2)}$ and $H_1^{(3)}$ follows the same rule (their entries can be different from $H_1^{(1)}$) and after adding $H_1^{(1)} + H_1^{(2)} + H_1^{(3)}$ we obtain

$$H_1 = \frac{1}{3} \begin{bmatrix} -1 & 2 & 2 & 0 & 0 & 0 & 0 & 0 & 0 \\ 2 & -1 & 2 & 0 & 0 & 0 & 0 & 0 & 0 \\ 2 & 2 & -1 & 0 & 0 & 0 & 0 & 0 & 0 \\ 0 & 0 & 0 & -1 & 2 & 2 & 0 & 0 & 0 \\ 0 & 0 & 0 & 2 & -1 & 2 & 0 & 0 & 0 \\ 0 & 0 & 0 & 2 & 2 & -1 & 0 & 0 & 0 \\ 0 & 0 & 0 & 0 & 0 & 0 & -1 & 2 & 2 \\ 0 & 0 & 0 & 0 & 0 & 0 & 2 & -1 & 2 \\ 0 & 0 & 0 & 0 & 0 & 0 & 2 & 2 & -1 \end{bmatrix}. \tag{5.25}$$

Section 5.3.1 describes another recipe for obtaining $H_1^{(1)}$ by associating a unit vector to each polygon. Using Eq. (5.22), we obtain

$$\left| P_1^1 \right\rangle = \frac{1}{\sqrt{3}} \left(|0\rangle + |1\rangle + |2\rangle \right),$$

$$\left| P_1^2 \right\rangle = \frac{1}{\sqrt{3}} \left(|3\rangle + |4\rangle + |5\rangle \right),$$

$$\left| P_1^3 \right\rangle = \frac{1}{\sqrt{3}} \left(|6\rangle + |7\rangle + |8\rangle \right).$$

Using Eq. (5.23) we obtain the same Hamiltonian H_1 of Eq. (5.25). Notice that $H_1^2 = I$, that is, H_1 is Hermitian and unitary. Besides, H_1 is a local operator. For example, if the walker is on vertex 0, the action of H_1 on $|0\rangle$ is a superposition of vertices in the neighborhood of 0, in fact,

$$H_1 |0\rangle = -\frac{1}{3} |0\rangle + \frac{2}{3} |1\rangle + \frac{2}{3} |2\rangle. \tag{5.26}$$

Notice that there is a nonzero probability of finding the walker on vertex 0, which the vertex from which the walker departed. This is a characteristic of the staggered model and only in very special cases the probability of staying is zero. The action of H_1 does not move the walker to all neighbors, for instance, vertices 7 and 8 are in the neighborhood of 0, but those vertices are absent from the sum on the right hand side of Eq. (5.26). This is also a characteristic of the staggered model, which is necessary in order to fulfill unitarity.

We keep following the recipe given in Sect. 5.3.1 in order to find Hamiltonian H_2, which is associated with tessellation $\mathscr{T}_2$. We must choose three Hermitian matrices $H_2^{(1)}$, $H_2^{(2)}$, $H_2^{(3)}$ because tessellation $\mathscr{T}_2$ has three polygons. Let us define $H_2^{(1)}$, which is associated with $P_2^{(1)} = \{0, 7, 8\}$, as

$$H_2^{(1)} = \frac{1}{3} \begin{bmatrix} -1 & 0 & 0 & 0 & 0 & 0 & 0 & 2 & 2 \\ 0 & 0 & 0 & 0 & 0 & 0 & 0 & 0 & 0 \\ 0 & 0 & 0 & 0 & 0 & 0 & 0 & 0 & 0 \\ 0 & 0 & 0 & 0 & 0 & 0 & 0 & 0 & 0 \\ 0 & 0 & 0 & 0 & 0 & 0 & 0 & 0 & 0 \\ 0 & 0 & 0 & 0 & 0 & 0 & 0 & 0 & 0 \\ 0 & 0 & 0 & 0 & 0 & 0 & 0 & 0 & 0 \\ 2 & 0 & 0 & 0 & 0 & 0 & 0 & -1 & 2 \\ 2 & 0 & 0 & 0 & 0 & 0 & 0 & 2 & -1 \end{bmatrix}. \tag{5.27}$$

Notice that $H_2^{(1)}$ acts non-trivially only on the entries associated with $P_2^{(1)}$. By choosing similar matrices for $H_2^{(2)}$ and $H_2^{(3)}$ and after adding $H_2^{(1)} + H_2^{(2)} + H_2^{(3)}$ we obtain

$$H_2 = \frac{1}{3} \begin{bmatrix} -1 & 0 & 0 & 0 & 0 & 0 & 0 & 2 & 2 \\ 0 & -1 & 2 & 2 & 0 & 0 & 0 & 0 & 0 \\ 0 & 2 & -1 & 2 & 0 & 0 & 0 & 0 & 0 \\ 0 & 2 & 2 & -1 & 0 & 0 & 0 & 0 & 0 \\ 0 & 0 & 0 & 0 & -1 & 2 & 2 & 0 & 0 \\ 0 & 0 & 0 & 0 & 2 & -1 & 2 & 0 & 0 \\ 0 & 0 & 0 & 0 & 2 & 2 & -1 & 0 & 0 \\ 2 & 0 & 0 & 0 & 0 & 0 & 0 & -1 & 2 \\ 2 & 0 & 0 & 0 & 0 & 0 & 0 & 2 & -1 \end{bmatrix}. \tag{5.28}$$

Using the second recipe based on Eq. (5.22) for obtaining $H_2^{(1)}$, we define vectors

$$\left| P_2^1 \right\rangle = \frac{1}{\sqrt{3}} \left(|0\rangle + |7\rangle + |8\rangle \right),$$

$$\left| P_2^2 \right\rangle = \frac{1}{\sqrt{3}} \left(|1\rangle + |2\rangle + |3\rangle \right),$$

$$\left| P_2^3 \right\rangle = \frac{1}{\sqrt{3}} \left(|4\rangle + |5\rangle + |6\rangle \right).$$

Using Eq. (5.23) we obtain the same Hamiltonian H_2 of Eq. (5.28). Notice again that $H_2^2 = I$, that is, H_2 is also Hermitian and unitary. H_2 is also local, but it is complementary to the locality of H_1 because if the walker is on vertex v, H_2 spreads the position of the walker over the neighbors of v that are in the blue polygon that contains v. Some neighbors of v are unreachable by the action of H_2. We have already shown that the same is true for H_1.

We have associated each tessellation with a local Hermitian operator. Now we have to go further in order to associate each tessellation with a local unitary operator. This is accomplished by employing Eq. (5.20) after choosing some angle θ. In this example, we have

$$U_1 = e^{i\theta H_1},$$
$$U_2 = e^{i\theta H_2}. \tag{5.29}$$

Since $H_1^2 = H_2^2 = I$, we also have

$$U_1 = \cos(\theta)\, I + i\sin(\theta)\, H_1,$$
$$U_2 = \cos(\theta)\, I + i\sin(\theta)\, H_2. \tag{5.30}$$

Notice that U_1 and U_2 are local operators because H_1 and H_2 are local operators and the action of $\cos(\theta)I$ (the first term of the expressions of U_1 and U_2) does not push the walker away from his neighborhood.

Finally, we are now able to define the evolution operator, which is

$$\mathcal{U} = U_2\, U_1. \tag{5.31}$$

As expected, $\mathcal{U}$ is not local. This means that, if the initial position is localized, the successive action of $\mathcal{U}$ will spread the walker's position over the graph vertices. This is one of the most desired and expected behavior from a quantum-walk evolution operator. For example, the action of $\mathcal{U}$ on $|0\rangle$ with $\theta = \pi/4$ yields

$$\mathcal{U}\,|0\rangle = \frac{1}{9}\Big((4 - 3i)\,|0\rangle - (1 - 3i)\,|1\rangle - (1 - 3i)\,|2\rangle - 4\,|3\rangle +$$
$$(1 + 3i)\,|7\rangle + (1 + 3i)\,|8\rangle \Big). \tag{5.32}$$

The action of $\mathcal{U}$ spreads the initial position over the clique that contains vertex 0 and the neighboring cliques. The neighboring cliques belong to a tessellation distinct from the one that contains the initial position. This alternation of tessellations is the only way to define non-trivial unitary evolution operators when time is discrete.

Exercise 5.10 Prove that if H_1 and H_2 are Hermitian then $H_1 + H_2$ is Hermitian. Prove that if U_1 and U_2 are unitary then $U_1 U_2$ is unitary. Give an example of two Hermitian matrices such that their product is not Hermitian. Give an example of two unitary matrices such that their sum is not unitary.

Exercise 5.11 Find the action of $U_1 U_2$ on $|0\rangle$ when $\theta = \pi/4$, where U_1 and U_2 are given by Eq. (5.29), and prove that U_1 and U_2 do not commute.

5.3.3 Diamond Ring of Size N

In this section, we address a diamond ring of size $N = 3n$ as depicted in Fig. 5.3. N is the number of vertices and n is the number of diamonds. The goal now is to obtain not only the evolution operator but also the walker's state at any time step t.

Fig. 5.3 A diamond ring of
size $N = 3n$ covered by two
tessellations

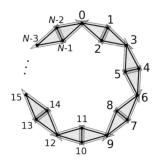

This is possible in this case because the diamond ring has a rotational symmetry,
which we explore using the Fourier basis.

The evolution operator is has the same form of Eq. (5.31), and the local unitary
operators are

$$U_1 = e^{i\theta H_1},$$
$$U_2 = e^{i\theta H_2},$$

where Hamiltonians H_1 and H_2 are given by

$$H_1 = 2 \sum_{\ell=0}^{n-1} \left| P_\ell^1 \right\rangle \left\langle P_\ell^1 \right| - I,$$

$$H_2 = 2 \sum_{\ell=0}^{n-1} \left| P_\ell^2 \right\rangle \left\langle P_\ell^2 \right| - I,$$

and the vectors associated with the polygons are

$$\left| P_\ell^1 \right\rangle = \frac{1}{\sqrt{3}} \left(|2\ell, 0\rangle + |2\ell + 1, 0\rangle + |2\ell + 1, 1\rangle \right),$$

$$\left| P_\ell^2 \right\rangle = \frac{1}{\sqrt{3}} \left(|2\ell + 1, 0\rangle + |2\ell + 1, 1\rangle + |2\ell + 2, 0\rangle \right).$$

The evolution operator $\mathcal{U}$ can be explicitly calculated from the above definitions.
For instance, taking $n = 3$ and $\theta = \pi/4$, we obtain[8]

---

[8] At this point, it is highly recommended some computer algebra system, such as Sage (for those
that prefer free software) or Mathematica or Maple.

$$\mathscr{U} = \frac{1}{9} \begin{bmatrix} 3+\bar{a} & a & a & 0 & 0 & 0 & -4 & -\bar{a} & -\bar{a} \\ -\bar{a} & 1+\bar{a} & 2a & a & -2 & -2 & 0 & 0 & 0 \\ -\bar{a} & 2a & 1+\bar{a} & a & -2 & -2 & 0 & 0 & 0 \\ -4 & -\bar{a} & -\bar{a} & 3+\bar{a} & a & a & 0 & 0 & 0 \\ 0 & 0 & 0 & -\bar{a} & 1+\bar{a} & 2a & a & -2 & -2 \\ 0 & 0 & 0 & -\bar{a} & 2a & 1+\bar{a} & a & -2 & -2 \\ 0 & 0 & 0 & -4 & -\bar{a} & -\bar{a} & 3+\bar{a} & a & a \\ a & -2 & -2 & 0 & 0 & 0 & -\bar{a} & 1+\bar{a} & 2a \\ a & -2 & -2 & 0 & 0 & 0 & -\bar{a} & 2a & 1+\bar{a} \end{bmatrix}, \qquad (5.33)$$

where $a = 1 + 3i$ and $\bar{a} = 1 - 3i$.

The goal now is to find the spectrum of $\mathscr{U}$ with the help of the Fourier basis. The standard recipe to obtain the Fourier basis in the staggered model is the following one: We have to check which vertices of the diamond ring has the same type using the symmetries of the graph, in this case, we use the rotational symmetry. By exploring in details Fig. 5.3, we convince ourselves that vertices $\{0, 3, 6, \ldots, N-3\}$ have the same type. Besides, there are two more classes: $\{1, 4, 7, \ldots, N-2\}$ and $\{2, 5, 8, \ldots, N-1\}$. Notice that there is an extra symmetry. Each diamond can flipped along the axis connecting vertices $(2j, 0)$ and $(2j + 2, 0)$ for $0 \le j < n$ without changing the picture. Next step is to use powers of the primitive nth root of unity

$$\omega = e^{\frac{\pi i}{n}}$$

as amplitudes of the following vectors:

$$\left| \psi_0^k \right\rangle = \frac{1}{\sqrt{n}} \sum_{\ell=0}^{n-1} \omega^{2\ell k} |2\ell, 0\rangle,$$

$$\left| \psi_1^k \right\rangle = \frac{1}{\sqrt{2n}} \sum_{\ell=0}^{n-1} \omega^{(2\ell+1)k} \left(|2\ell + 1, 0\rangle + |2\ell + 1, 1\rangle \right), \qquad (5.34)$$

$$\left| \psi_2^k \right\rangle = \frac{1}{\sqrt{2n}} \sum_{\ell=0}^{n-1} \omega^{(2\ell+1)k} \left(|2\ell + 1, 0\rangle - |2\ell + 1, 1\rangle \right).$$

The set $\left\{ \left| \psi_j^k \right\rangle : 0 \le j \le 2, \ 0 \le k < N \right\}$ is an orthogonal basis of the Hilbert space $\mathscr{H}^N$, that is, there are N vectors $\left| \psi_j^k \right\rangle$, which obey $\langle \psi_{j'}^{k'} | \psi_j^k \rangle = \delta_{jj'} \delta_{kk'}$.

This basis is called Fourier basis and has the following property: If we fix k, vectors $\left| \psi_0^k \right\rangle$, $\left| \psi_1^k \right\rangle$, and $\left| \psi_2^k \right\rangle$ define a 3-dimensional subspace of the Hilbert space that is invariant under the action of the evolution operator. This means that

$$\mathscr{U}\left|\psi_0^k\right\rangle = u_{00}\left|\psi_0^k\right\rangle + u_{10}\left|\psi_1^k\right\rangle + u_{20}\left|\psi_2^k\right\rangle,$$

$$\mathscr{U}\left|\psi_1^k\right\rangle = u_{01}\left|\psi_0^k\right\rangle + u_{11}\left|\psi_1^k\right\rangle + u_{21}\left|\psi_2^k\right\rangle, \tag{5.35}$$

$$\mathscr{U}\left|\psi_2^k\right\rangle = u_{02}\left|\psi_0^k\right\rangle + u_{12}\left|\psi_1^k\right\rangle + u_{22}\left|\psi_2^k\right\rangle,$$

where coefficients u_{ij} for $0 \le i, j \le 2$ are complex numbers that form a 3-dimensional matrix, which we call $\mathscr{U}_{\text{RED}}$ (reduced $\mathscr{U}$). We can calculate those coefficients when $\theta = \pi/4$ and eventually we obtain

$$\mathscr{U}_{\text{RED}} = \begin{bmatrix} a_k & b_k & 0 \\ b_k & \bar{a}_k & 0 \\ 0 & 0 & -i \end{bmatrix}, \tag{5.36}$$

where

$$a_k = \frac{4 - 3i}{9} - \frac{4}{9}\omega^{2k},$$

$$b_k = \frac{2\sqrt{2}\,i}{9}\left(3\cos\frac{k\pi}{n} - \sin\frac{k\pi}{n}\right),$$

and $\bar{a}_k$ is the complex conjugate of a_k. Notice that $\mathscr{U}_{\text{RED}}\,\mathscr{U}_{\text{RED}}^\dagger = I$, that is, $\mathscr{U}_{\text{RED}}$ is unitary, because $|a_k|^2 + |b_k|^2 = 1$ and $b_k + \bar{b}_k = 0$.

If a graph family had no translational or rotational symmetry, an *Anzatz* similar to the one given by Eq. (5.35) would not work, and the calculation of the spectrum of the evolution operator would be burdensome or even unattainable for the family.

Notice that $\mathscr{U}_{\text{RED}}$ is the direct sum of a 2×2 matrix with a 1×1 matrix. Then, the spectrum of $\mathscr{U}_{\text{RED}}$ is obtained from the spectrum of each matrix. The 2×2 matrix has eigenvalues $\exp(\pm i\phi_k)$, where

$$\cos\phi_k = \frac{4}{9}\left(1 - \cos\frac{6\pi k}{N}\right). \tag{5.37}$$

The eigenvectors are

$$|v_0\rangle = \frac{1}{\sqrt{c_k^+}}\begin{bmatrix} b_k \\ e^{i\phi_k} - a_k \\ 0 \end{bmatrix} \quad \text{and} \quad |v_1\rangle = \frac{1}{\sqrt{c_k^-}}\begin{bmatrix} b_k \\ e^{-i\phi_k} - a_k \\ 0 \end{bmatrix}, \tag{5.38}$$

where

$$c_k^\pm = 2\left(1 - \Re(a_k)\cos\phi_k \mp \Im(a_k)\sin\phi_k\right), \tag{5.39}$$

$\Re(a_k)$ and $\Im(a_k)$ are the real and imaginary parts of a_k, respectively. The 1×1 matrix has eigenvalue $-i$ and eigenvector

$$|v_2\rangle = \begin{bmatrix} 0 \\ 0 \\ 1 \end{bmatrix}. \tag{5.40}$$

The eigenvalues of $\mathscr{U}$ for a diamond ring of any size are $\exp(\pm i\phi_k)$ and $-i$. The eigenvectors are

$$\left|V_\ell^k\right\rangle = \sum_{j=0}^{2} \langle j|v_\ell\rangle \left|\psi_j^k\right\rangle, \tag{5.41}$$

where $0 \le k < n$ and $0 \le \ell \le 2$. Notice that $\left|V_2^k\right\rangle = \left|\psi_2^k\right\rangle$.

Now we can obtain the quantum-walk state $|\psi(t)\rangle$ after t time steps assuming that the initial state is $|\psi(0)\rangle$. The master equation is

$$|\psi(t)\rangle = \mathscr{U}^t |\psi(0)\rangle. \tag{5.42}$$

Using the spectral decomposition of $\mathscr{U}$, we write

$$\mathscr{U}^t = \sum_{\ell=0}^{2} \sum_{k=0}^{n-1} (\lambda_\ell^k)^t \left|V_\ell^k\right\rangle\left\langle V_\ell^k\right|, \tag{5.43}$$

where λ_ℓ^k is the eigenvalue associated with $\left|V_\ell^k\right\rangle$.

Let us assume that the walk departs from the origin $|\psi(0)\rangle = |0,0\rangle$. Using Eqs. (5.41) and (5.34), we obtain

$$\langle V_\ell^k|\psi(0)\rangle = \frac{1}{\sqrt{n}} \langle v_\ell|0\rangle, \tag{5.44}$$

Using the last three equations, we obtain

$$|\psi(t)\rangle = \frac{1}{\sqrt{n}} \sum_{k=0}^{n-1} \bar{b}_k \left(\frac{e^{i\phi_k t}}{\sqrt{c_k^+}} \left|V_0^k\right\rangle + \frac{e^{-i\phi_k t}}{\sqrt{c_k^-}} \left|V_1^k\right\rangle \right). \tag{5.45}$$

Exercise 5.12 The goal of this exercise is to obtain matrix $\mathscr{U}_{\mathrm{RED}}$ of Eq. (5.36). First, find the reduced form of matrices H_1 and H_2. Second, find the reduced form of matrices U_1 and U_2 using Eq. (5.29). Finally, $\mathscr{U}_{\mathrm{RED}}$ is the product of reduced form of matrices U_1 and U_2 in the proper order.

Exercise 5.13 Show that $\sum_{\ell=0}^{N-1} \omega^\ell = 0$. Show that $\{\left|\psi_j^k\right\rangle : 0 \leq j \leq 2, 0 \leq k < N\}$ is an orthogonal basis of the Hilbert space $\mathscr{H}^N$.

Exercise 5.14 Show that the eigenvalues of the reduced matrix u are exactly the same eigenvalues of $\mathscr{U}$. Show that the eigenvectors of $\mathscr{U}$ are the ones described by Eqs. (5.38) to (5.40). It seems a contradiction that the spectrum of $\mathscr{U}$ can be calculated from a 3-dimensional matrix; show that the number of eigenvalues and eigenvectors obtained from u when we consider all values of k matches exactly the number of eigenvalues and eigenvectors of $\mathscr{U}$.

5.3.4 Reproducing the Flip-Flop Coined Model

In this subsection, we show that the flip-flop coined model can be seen as a two-tessellable staggered quantum walk, that is, both models have exactly the same evolution operator. This connection helps to re-interpret the coined model and shows that the coin and shift operators have the same nature.

Consider a flip-flop coined quantum walk on a graph $\Gamma(V, E),$[9] where V is the vertex set and E is the edge set. Let $|V|$ be the number of vertices and $|E|$ the number of edges. Let $0, \ldots, |V|-1$ be the vertex labels and $0, \ldots, |E|-1$ the edge labels. The shift operator S of the flip-flop model is defined as

$$S|v, e\rangle = |v', e\rangle, \tag{5.46}$$

where v and v' are adjacent vertices and $e = (v, v')$ is an edge that connects v and v'. The shift operator belongs to the Hilbert space $\mathscr{H}^{2|E|}$. The dimension of the Hilbert space is $2|E|$ because for each edge e there are two basis vectors: $|v, e\rangle$ and $|v', e\rangle$, where v and v' are the vertices incident to e. The computational basis is the union of sets $\{|v, e\rangle, |v', e\rangle\}$ for all $e \in E$. Notice that we cannot split $|v, e\rangle$ as $|v\rangle \otimes |e\rangle$ unless the graph is regular.

Exercise 5.15 Show that if Γ is a d-regular graph then the dimension of the Hilbert space of the coined model is $d |V|$.

Exercise 5.16 Show that if S is defined by Eq. (5.46) then $S^2 = I$.

Exercise 5.17 Find the matrix representation of S for the complete graph of four vertices. Show that there is an ordering for the computational basis such that $S = I_6 \otimes X$.

The evolution operator of the flip-flop coined quantum walk is

$$U_{\text{coined}} = S C, \tag{5.47}$$

---

[9] In this context, Γ can be a multigraph.

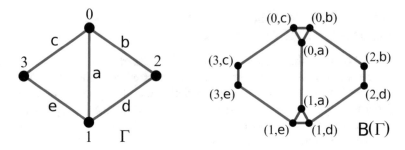

Fig. 5.4 A diamond Γ (left-hand graph) and its blow-up graph $B(\Gamma)$ (right-hand graph). Any flip-flop coined model defined on Γ can be reproduced by a 2-tessellable staggered quantum walk on $B(\Gamma)$ with the tessellations shown in blue and red colors

where C is the coin operator, which also belongs to the Hilbert space $\mathscr{H}^{2|E|}$. If graph Γ is d regular, it is possible to choose C so that it can be factored as $I \otimes G$, where G is a $d \times d$ matrix. In this case, coin G is applied to all vertices. In the general case, C is a little bit harder to define. We show an example after defining the notion of *blow-up* graph [7, 9]

The *blow-up* of a graph Γ is a new simple graph $B(\Gamma)$ that is obtained by replacing each vertex of Γ by a clique, whose size is the degree of the vertex, keeping the same neighborhood. Instead of giving a formal definition, let us explain this concept using the example in Fig. 5.4, which shows the blow-up of a diamond graph. The blow-up is a two-step procedure. In the first step, each vertex v of the original graph is converted into a clique, the size of which is the degree of vertex v in Γ. In Fig. 5.4, there are two degree-3 vertices and two degree-2 vertices in the original graph. In the second step, each new clique merges with the original graph by defining their neighborhood. The reverse process is also well defined: each maximal clique shrinks to a single vertex keeping the edges of the original graph. The reduced graph can be a multigraph, for instance, when two maximal cliques share more than one vertex.

Now we show an example that explain how a flip-flop coined quantum walk on a graph Γ is equivalent to a 2-tessellable staggered quantum walk on the blow-up graph $B(\Gamma)$. The meaning of equivalence in this context is that the evolution operators generated by each model are exactly the same.

Figure 5.4 shows a diamond graph Γ (left-hand graph) with vertices $\{0, 1, 2, 3\}$ and edges $\{a, b, c, d, e\}$. The Hilbert space is spanned by $\mathscr{H} = \text{span}\{|0, a\rangle, |0, b\rangle, |0, c\rangle, |1, a\rangle, |1, d\rangle, |1, e\rangle, |2, b\rangle, |2, d\rangle, |3, c\rangle, |3, e\rangle\}$. The shift operator is given by Eq. (5.46). For instance, $S|0, b\rangle = |2, b\rangle$. Let us consider the following 10×10 coin operator

$$C = \begin{bmatrix} G & 0 & 0 & 0 \\ 0 & G & 0 & 0 \\ 0 & 0 & H & 0 \\ 0 & 0 & 0 & H \end{bmatrix}, \tag{5.48}$$

where where H is the Hadamard gate and G is the 3×3 Grover coin

$$G = \frac{1}{3} \begin{bmatrix} -1 & 2 & 2 \\ 2 & -1 & 2 \\ 2 & 2 & -1 \end{bmatrix}. \tag{5.49}$$

Now we write the evolution operator for the staggered model on the blow-up graph $B(\Gamma)$ depicted as the right-hand graph in Fig. 5.4. The evolution operator is

$$U_{\text{stagg}} = U_{\text{red}} \, U_{\text{blue}},$$

where U_{red} and U_{blue} are obtained from the red and blue tessellations. Notice that there is a labeling rule for $B(\Gamma)$. When we blow up vertex 0 of Γ, there are three new vertices in $B(\Gamma)$, which receives labels $(0, a)$, $(0, b)$, and $(0, c)$, because vertex 0 is incident to edges a, b, and c of Γ. The same rule applies to the other blown-up vertices. The Hilbert space for the staggered model is spanned by the same basis of the Hilbert space of the coined model.

Let us obtain U_{blue} using the staggered model. We choose the following vectors for each polygon:

$$|P_0\rangle = \frac{|0, a\rangle + |0, b\rangle + |0, c\rangle}{\sqrt{3}},$$

$$|P_1\rangle = \frac{|1, a\rangle + |1, d\rangle + |1, e\rangle}{\sqrt{3}},$$

$$|P_2\rangle = \frac{\sqrt{2 + \sqrt{2}}\,|2, b\rangle + \sqrt{2 - \sqrt{2}}\,|2, d\rangle}{2},$$

$$|P_3\rangle = \frac{\sqrt{2 + \sqrt{2}}\,|3, c\rangle + \sqrt{2 - \sqrt{2}}\,|3, e\rangle}{2}.$$

Using Eq. (5.23) we have

$$H_{\text{blue}} = 2 \sum_{\ell=0}^{3} |P_\ell\rangle\langle P_\ell| - I. \tag{5.50}$$

It is straightforward to check that $U_{\text{blue}} = \exp(i\theta \, H_{\text{blue}})$ is equal to matrix (5.48) when $\theta = \pi/2$ modulo a global phase.

Let us obtain U_{red} using the staggered model. We choose the following vectors for each polygon, using the edges as labels:

$$|P_a\rangle = \frac{|0, a\rangle + |1, a\rangle}{\sqrt{2}},$$

$$|P_b\rangle = \frac{|0, b\rangle + |2, b\rangle}{\sqrt{2}},$$

$$|P_c\rangle = \frac{|0, c\rangle + |3, c\rangle}{\sqrt{2}},$$

$$|P_d\rangle = \frac{|1, d\rangle + |2, d\rangle}{\sqrt{2}},$$

$$|P_e\rangle = \frac{|1, e\rangle + |3, e\rangle}{\sqrt{2}}.$$

Using Eq. (5.23) we have

$$H_{\text{red}} = 2 \sum_{\ell=a}^{e} |P_\ell\rangle\langle P_\ell| - I. \tag{5.51}$$

Now let us check whether $H_{\text{red}} = S$. Let us start by applying H_{red} on the first vector of the computational basis:

$$H_{\text{red}}|0, a\rangle = 2 \sum_{\ell=a}^{e} |P_\ell\rangle\langle P_\ell|0, a\rangle - |0, a\rangle$$

$$= \frac{2}{\sqrt{2}}|P_a\rangle - |0, a\rangle$$

$$= |1, a\rangle.$$

After applying H_{red} on all vectors of the computational basis, we convince ourselves that $H_{\text{red}} = S$. Define $U_{\text{red}} = \exp(i\pi/2\, H_{\text{red}})$, which is equal to H_{red} modulo a global phase. We have just shown that $U_{\text{coined}} = U_{\text{stagg}}$.

References

1. Abreu, A., Cunha, L., Fernandes, T., de Figueiredo, C., Kowada, L., Marquezino, F., Posner, D., Portugal, R.: The graph tessellation cover number: extremal bounds, efficient algorithms and hardness. In: Bender, M.A., Farach-Colton, M., Mosteiro, M.A. (eds.) LATIN 2018: Theoretical Informatics, pp. 1–13. Springer, Cham (2018)
2. Aharonov, Y., Davidovich, L., Zagury, N.: Quantum random walks. Phys. Rev. A **48**, 1687–1690 (1993). https://doi.org/10.1103/PhysRevA.48.1687

3. Konno, N., Portugal, R., Sato, I., Segawa, E.: Partition-based discrete-time quantum walks. Quantum Inf. Process. **17**(4), 100 (2018)
4. Lawler, G.F., Limic, V.: Random Walk: A Modern Introduction, 1st edn. In: Cambridge Studies in Advanced Mathematics. Cambridge University Press, Cambridge (2010)
5. Meyer, D.A.: From quantum cellular automata to quantum lattice gases. J. Stat. Phys. **85**(5), 551–574 (1996). https://doi.org/10.1007/BF02199356
6. Philipp, P., Portugal, R.: Exact simulation of coined quantum walks with the continuous-time model. Quantum Inf. Proc. **16**(1), 14 (2017)
7. Portugal, R.: Staggered quantum walks on graphs. Phys. Rev. A **93**, 062,335 (2016)
8. Portugal, R., Santos, R.A.M., Fernandes, T.D., Gonçalves, D.N.: The staggered quantum walk model. Quantum Inf. Proc. **15**(1), 85–101 (2016)
9. Portugal, R., de Oliveira, M.C., Moqadam, J.K.: Staggered quantum walks with Hamiltonians. Phys. Rev. A **95**, 012,328 (2017)
10. Spitzer, F.: Principles of Random Walk, 2 edn. In: Graduate Texts in Mathematics. Springer, New York (1964)
11. Vizing, V.G.: On an estimate of the chromatic class of a p-graph. Discret. Analiz. **3**, 25–30 (1964)

Chapter 6
Conclusion and Further Remarks

*"I seem to have been only like a boy playing on the seashore,
and diverting myself in now and then finding a smoother pebble
or a prettier shell than ordinary, whilst the great ocean of truth
lay all undiscovered before me." (Isaac Newton)*

Quantum computing is a fascinating field of research which initiated in the early 1980s and is continuously increasing in importance since then. Many algorithms have been developed, some of them with exponential speedup when compared to the best known classical algorithms. Even the hardware realizations, which represented a great technological challenge in the early years of quantum computing, is now advancing much more quickly than expected. Being prepared to deal with this new technology in the near future is crucial for students of science, technology, engineering and mathematics.

Nevertheless, the study of quantum computing is a laborious undertaking due to its inherent interdisciplinarity. Students of computer science and mathematics often feel intimidated by concepts of quantum mechanics, and similarly students of physics and engineering tend to have more difficulties with some concepts of theoretical computer science.

In the present book, we could not possibly cover all relevant topics of quantum computing, since it would require a massive amount of pages. Instead, we decided to focus on a common core of concepts in quantum computing. The goal was to provide the reader with all the elements—such as definitions, notations and important algorithms—necessary to later investigate more specialized themes. Some suggestions of topics for further studies include the following:

- **Quantum error correction codes.** Both classical and quantum logic gates are subject to errors of different kinds. Therefore, in order to perform reliable computations it is fundamental to have means of detecting and correcting errors. Fortunately, since middle 1990s several quantum error correction codes have been designed [19]. This is still a very active research line, with many interesting open problems. For details, the reader may refer to Lidar and Brun [17].

© The Author(s), under exclusive license to Springer Nature Switzerland AG 2019
F. de Lima Marquezino et al., *A Primer on Quantum Computing*, SpringerBriefs in
Computer Science, https://doi.org/10.1007/978-3-030-19066-8_6

- **Quantum key exchange and post-quantum cryptography.** Some of the quantum algorithms with exponential speedup compared to their classical counterparts can be directly applied to code breaking. Therefore, we may expect that together with the development of reliable quantum hardware, some of the classical protocols for cryptography will become obsolete. There are two types of solution for that problem. The first one is quantum key distribution, a family of protocols that use quantum mechanics to implement a perfectly secure key exchange protocol [6]. The second one is post-quantum cryptography, a family of classical protocols of cryptography that are resistant against attacks of quantum computers. Both approaches have very active communities and several open problems.

- **Adiabatic algorithms.** The adiabatic quantum computing model is based on the adiabatic theorem, which states that a quantum system remains in its instantaneous eigenstate if it evolves slowly enough, and if there is a gap between its corresponding eigenvalue and the rest of the Hamiltonian's spectrum [9]. This model of computing seeks to design algorithms by choosing two Hamiltonians: one for which the ground state is easy to prepare, and one for which the ground state encodes the solution for the problem. The quantum algorithm evolves according to an interpolation of those two Hamiltonians. This approach is usually applied to optimization problems, in a process analogous to the simulated annealing—and, for that reason, is also known as *quantum annealing*.

- **Optimization algorithms.** Optimization is one of the most promising areas where quantum computers may help to achieve better results than their classical counterparts [15]. Grover's algorithm, for example, can be used as the basis for discrete optimization algorithms [4, 11]. The quantum approximate optimization algorithm (QAOA) is a more recent approach [12]. Quantum computers can be used to solve systems of linear equations [14], which in turn can be used as the basis for a quantum algorithm for the data fitting problem [20], for example.

- **Quantum machine learning.** Many problems in machine learning reduce to optimization problems. Since optimization is one of the main applications of quantum computing, it seems natural that machine learning may also benefit from quantum algorithms. In fact, a lot of progress has been made in quantum machine learning in the last few years, to the point that it is now considered one of the main applications of quantum computing [8, 18]. Since it is a relatively new subject, there are still several open problems and opportunities for research.

- **Quantum complexity.** Quantum computing is a source of many interesting problems for theoretical computer science [1, 7, 13]. It is imperative to investigate what can or cannot be efficiently computed by a quantum machine, and what consequences these results may have to classical computing. Finding non-trivial quantum lower bounds is also very important in practice, since it helps to determine which computational problems still admit better solutions [2, 3, 5].

- **Experimental realizations of quantum computing.** There are several proposals of techniques for building reliable quantum computers, each of them with their own advantages and disadvantages, and it is not yet clear which technology will become the industry standard. Therefore, the area of experimental quantum computing is still very active and with several important research problems [10, 16].

References

1. Aaronson, S.: Quantum Computing since Democritus. Cambridge University Press, New York (2013)
2. Ambainis, A.: Quantum lower bounds by quantum arguments. J. Comput. Syst. Sci. **64**(4), 750–767 (2002)
3. Ambainis, A.: Quantum query algorithms and lower bounds. In: Löwe, B., Piwinger, B., Räsch, T. (eds.) Classical and New Paradigms of Computation and their Complexity Hierarchies, pp. 15–32. Springer, Amsterdam (2004)
4. Baritompa, W.P., Bulger, D.W., Wood, G.R.: Grover's quantum algorithm applied to global optimization. SIAM J. Optim. **15**(4), 1170–1184 (2005)
5. Beals, R., Buhrman, H., Cleve, R., Mosca, M., De Wolf, R.: Quantum lower bounds by polynomials. J. ACM (JACM) **48**(4), 778–797 (2001)
6. Bennett, C.H., Brassard, G.: An update on quantum cryptography. In: Blakley, G.R., Chaum, D. (eds.) Advances in Cryptology, pp. 475–480. Springer Berlin Heidelberg, Berlin (1985)
7. Bernstein, E., Vazirani, U.: Quantum complexity theory. SIAM J. Comput. **26**(5), 1411–1473 (1997). https://doi.org/10.1137/S0097539796300921
8. Biamonte, J., Wittek, P., Pancotti, N., Rebentrost, P., Wiebe, N., Lloyd, S.: Quantum machine learning. Nature **549**(7671), 195 (2017)
9. Born, M., Fock, V.: Beweis des Adiabatensatze. Zeitschrift für Physik **51**(3), 165–180 (1928). https://doi.org/10.1007/BF01343193
10. Devoret, M.H., Schoelkopf, R.J.: Superconducting circuits for quantum information: An outlook. Science **339**(6124), 1169–1174 (2013). https://doi.org/10.1126/science.1231930
11. Durr, C., Hoyer, P.: A quantum algorithm for finding the minimum. ArXiv preprint quant-ph/9607014 (1996)
12. Farhi, E., Goldstone, J., Gutmann, S.: A quantum approximate optimization algorithm. Technical Report, Massachusetts Institute of Technology (2014). MIT-CTP/4610
13. Grillo, S., Marquezino, F.: Quantum query as a state decomposition. Theor. Comput. Sci. **736**, 62–75 (2018). https://doi.org/10.1016/j.tcs.2018.03.017
14. Harrow, A.W., Hassidim, A., Lloyd, S.: Quantum algorithm for linear systems of equations. Phys. Rev. Lett. **103**, 150,502 (2009). https://doi.org/10.1103/PhysRevLett.103.150502
15. Hogg, T., Portnov, D.: Quantum optimization. Inf. Sci. **128**(3), 181–197 (2000). https://doi.org/10.1016/S0020-0255(00)00052-9
16. Ladd, T.D., Jelezko, F., Laflamme, R., Nakamura, Y., Monroe, C., O'Brien, J.L.: Quantum computers. Nature **464**(45) (2010). https://doi.org/10.1038/nature08812
17. Lidar, D.A., Brun, T.A. (eds.): Quantum Error Correction. Cambridge University, Cambridge (2013)
18. Schuld, M., Sinayskiy, I., Petruccione, F.: An introduction to quantum machine learning. Contemp. Phys. **56**(2), 172–185 (2015). https://doi.org/10.1080/00107514.2014.964942
19. Shor, P.W.: Scheme for reducing decoherence in quantum computer memory. Phys. Rev. A **52**, R2493–R2496 (1995). https://doi.org/10.1103/PhysRevA.52.R2493
20. Wiebe, N., Braun, D., Lloyd, S.: Quantum algorithm for data fitting. Phys. Rev. Lett. **109**(5), 050,505 (2012)

Index

© The Author(s), under exclusive license to Springer Nature Switzerland AG 2019
F. de Lima Marquezino et al., *A Primer on Quantum Computing*, SpringerBriefs in
Computer Science, https://doi.org/10.1007/978-3-030-19066-8